Solutions Manual for

The Dynamics of Heat

Springer
New York
Berlin
Heidelberg
Barcelona
Budapest
Hong Kong
London
Milan
Paris
Santa Clara
Singapore
Tokyo

Hans U. Fuchs

Solutions Manual for The Dynamics of Heat

Springer

Hans U. Fuchs
Department of Physics
Technikum Winterhur
CH-8401 Winterhur
Switzerland

Printed on acid-free paper

Softcover reprint of the hardcover 1st edition 1996

Production managed by Karina Gershkovich; manufacturing supervised by Johanna Tschebull.
Photocomposed copy provided from the author's files.

9 8 7 6 5 4 3 2 1

ISBN-13: 978-0-387-94869-0 e-ISBN-13: 978-1-4612-2420-4

DOI: 10.1007/ 978-1-4612-2420-4

Preface

This manual contains detailed solutions of slightly more than half of the end of chapter problems in *The Dynamics of Heat*. The numbers of the problems included here are listed on the following page.

A friend who knows me well noticed that I have included only those problems which I could actually solve myself. Also, to make things more interesting, I have built random errors into the solutions. If you find any of them, please let me know. Also, if you have different ways of solving a problem, I would be happy to hear from you. Any feedback, also on the book in general, would be greatly appreciated.

There is an Errata sheet for the first printing of *The Dynamics of Heat*. By the time you read this, it should be available on the Internet for you to download. A reference to the URL of the sheet can be found in the announcement of my book on Springer's WWW pages (www.springer-ny.com).

Winterthur, 1996 *Hans Fuchs*

Numbers of Problems Solved

Prologue

1, 2, 4, 5, 6, 8, 12, 13, 17, 19, 23, 25, 27, 30, 32, 33, 34, 38, 39, 40, 42, 44, 47, 49, 50, 53, 55, 60, 61, 62

Chapter 1

2, 4, 5, 8, 9, 11, 13, 15, 16, 17, 18, 20, 21, 24, 26, 27, 29, 31, 33, 34, 37, 39, 41, 42, 44, 45, 47, 49, 51, 53, 55, 57, 58, 60, 62

Chapter 2

1, 3, 5, 6, 7, 9, 10, 12, 14, 15, 16, 17, 19, 20, 22, 23, 24, 26, 27, 29, 30, 32, 33, 36, 37, 38, 41, 42, 46, 47, 49

Interlude

2, 3, 4, 5, 6, 8, 10, 11, 12, 13, 18, 19, 20, 21, 23, 24, 28

Chapter 3

2, 4, 6, 8, 10, 12, 15, 16, 17, 18, 22, 24, 25, 28, 30, 31, 35, 36

Chapter 4

1, 2, 4, 6, 8, 9, 11, 12, 13, 15, 18, 20, 21, 22, 25, 27, 28, 29, 30, 31, 33, 34, 35, 39, 40, 43, 44, 46

Epilogue

1, 2, 11

PROLOGUE

Solutions of Selected Problems

Calculate the hydraulic capacitance of a glass tube used in a mercury pressure gauge. The inner diameter of the tube is 8.0 mm.

SOLUTION:

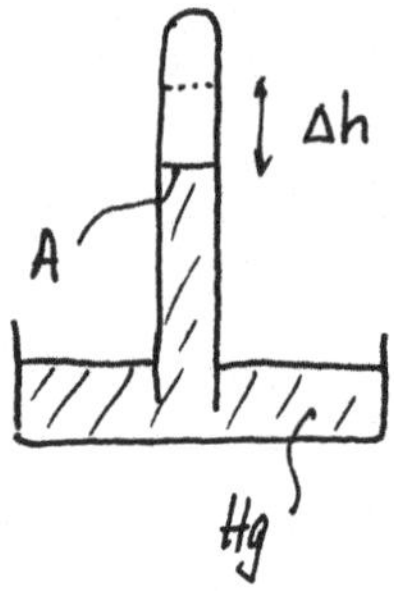

The hydraulic capacitance has been defined by Equation (9):

$$K = \frac{\Delta V}{\Delta P} = \frac{A \cdot \Delta h}{\rho_{Hg}\, g\, \Delta h} = \frac{A}{\rho_{Hg}\, g}$$

$$K = \frac{\pi\,(0.0040)^2}{13600 \cdot 9.81}\ \frac{m^3}{Pa} = 3.77 \cdot 10^{-10}\ \frac{m^3}{Pa}$$

The small value of K means that a small additional amount of mercury causes a large increase of pressure.

■

Derive the expression for the hydraulic capacitance of the conical container shown in Figure 21. Do the same for a U-tube.

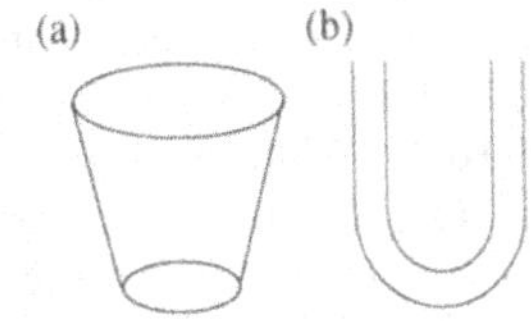

FIGURE 21. Problem 2.

SOLUTION:

For variable geometry containers, the definition of capacitance has to be adjusted:

a)

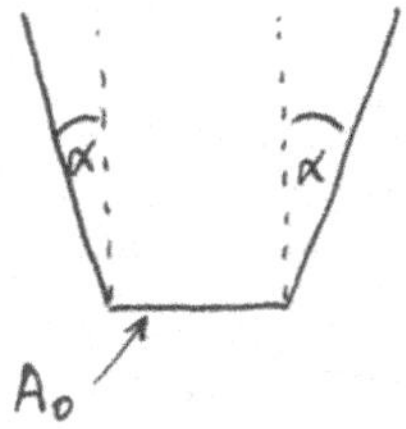

$$\dot{V} = K\dot{P} \quad \text{or} \quad K = \frac{dV}{dP}$$

Calculate the volume of fluid in the container as a function of pressure at the bottom and take the derivative:

$$V(h) = \int_0^h A(h)\,dh = \int_0^P A(h)\frac{1}{\rho g}\,dP \quad \Rightarrow \quad \frac{dV}{dP} = \frac{1}{\rho g}A(h)$$

Here: $A(h) = \pi\left(r_0 + tg\,\alpha \cdot h\right)^2$.

b)

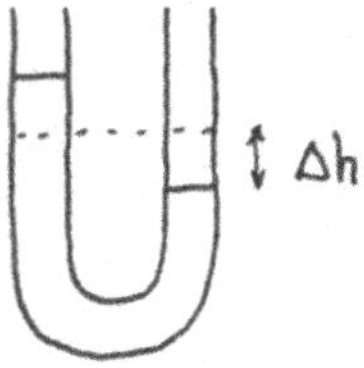

$$K = \frac{\Delta V}{\Delta P} = \frac{A \cdot \Delta h}{\rho g \cdot 2\Delta h} = \frac{1}{2}\frac{A}{\rho g}$$

We get the same result, i.e. half the capacitance of a glass tube (Problem 1), by calculating the capacitance of two equal containers connected in series.

■

Two currents of water are flowing into a fountain. The first changes linearly from 2.0 liters/s to 1.0 liters/s within the first 10 s. The second has a constant magnitude of 0.50 liters/s. In the time span from the beginning of the 4th second to the end of the 6th second, the volume of the water in the fountain decreases by 0.030 m^3. a) Calculate the volume flux of the current leaving the fountain. b) How much water will be in the fountain after 10 s, if the initial volume is equal to 200 liters?

SOLUTION :

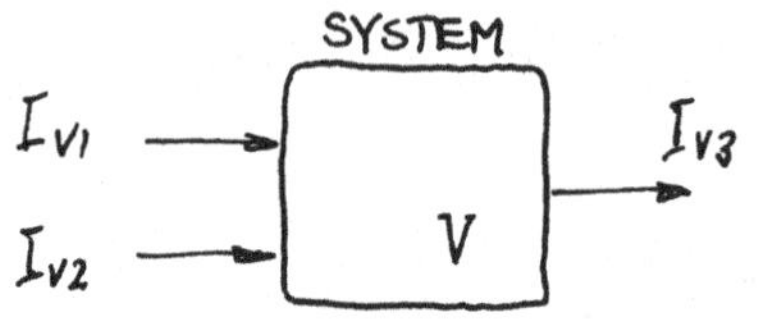

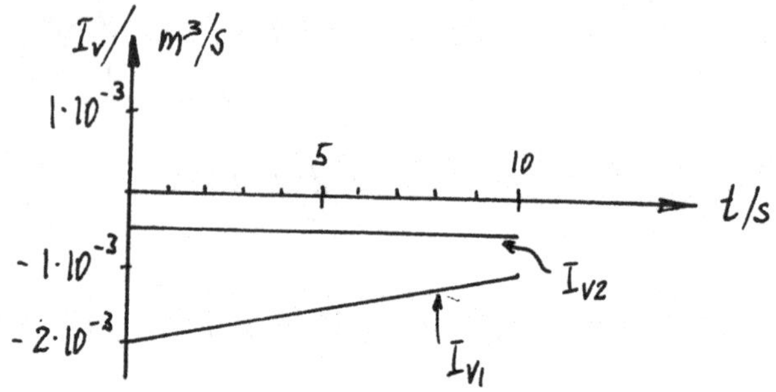

Balance of volume: $\dot{V} = -(I_{V1} + I_{V2} + I_{V3})$. Remember that fluxes of currents flowing into a system are counted as negative quantities. Assume $\dot{V}$ = const. during the first 10 s:

$$\dot{V} = \Delta V/\Delta t = \frac{-0.030\ m^3}{(6-3)s} = -0.010\ m^3/s$$

a)
$$I_{V3} = -\dot{V} - (I_{V1} + I_{V2})$$
$$= 0.010\ m^3/s - \left(-2.0\cdot10^{-3}\ m^3/s + 10^{-4}\ m^3/s^2 \cdot t - 0.5\cdot10^{-3}\ m^3/s\right)$$
$$= 0.0125\ m^3/s - 10^{-4}\ m^3/s^2 \cdot t$$

b)
$$\Delta V = V_f - V_i \qquad (f: \text{final},\ i: \text{initial})$$
$$V_f = V_i + \Delta V = V_i + \dot{V}\cdot\Delta t$$
$$= 0.20\ m^3 + (-0.010\ m^3/s)\cdot 10s$$
$$= 0.10\ m^3 .$$

■

Two containers are joined by a pipe as in Figure 22. The second container has both an inlet and an outlet. Assume the flow through the pipes to obey the law of Hagen and Poiseuille. a) Write the equations of balance of volume for the fluid in the containers. b) Derive the relation between volume of fluid and pressure of fluid at the bottom of each of the containers. c) Write the laws for the volume fluxes through both pipes. d) Derive the differential equations for the height of the fluid in each of the containers in terms of the hydraulic capacitance and resistance of the elements of the system.

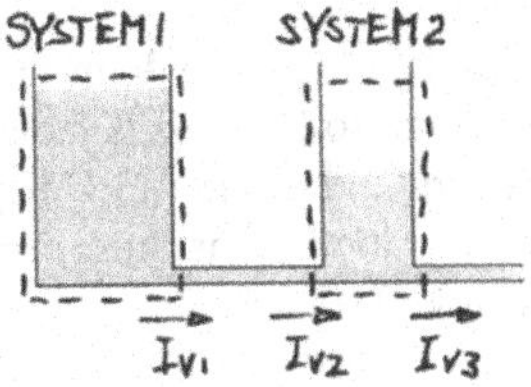

FIGURE 22. Problem 5.

SOLUTION:

a) System 1: $\dot{V}_1 = -I_{V1}$

System 2: $\dot{V}_2 = -(I_{V2} + I_{V3})$

b) Written in instantaneous form, using the capacitances:

$$\dot{V}_1 = K_1 \dot{P}_1 \quad , \quad \dot{V}_2 = K_2 \dot{P}_2$$

c) $$I_{V1} = -\frac{1}{R_1}\Delta P = -\frac{\pi r_1^4}{8\eta l_1}(P_2 - P_1)$$

$$I_{V2} = -I_{V1}$$

$$I_{V3} = -\frac{1}{R_2}\Delta P = -\frac{\pi r_2^2}{8\eta l_2}(P_a - P_2)$$, P_a: ambient pressure

d) Insert b and c in the equations of balance:

$$K_1 \dot{P}_1 = \frac{1}{R_1}(P_2 - P_1) \quad ; \quad K_2 \dot{P}_2 = -\frac{1}{R_1}(P_2 - P_1) + \frac{1}{R_2}(P_a - P_2)$$

$$\Rightarrow K_1 \dot{h}_1 = \frac{1}{R_1}(h_2 - h_1) \quad ; \quad K_2 \dot{h}_2 = -\frac{1}{R_1}(h_2 - h_1) - \frac{1}{R_2} h_2$$

Viscous oil is to be pumped from a shallow container into one lying 10 m higher up. The pipe has a diameter of 5.0 cm and a length of 20 m. If the mass flux is required to be 10 kg/s, how large should the power of the pump be? Draw the energy and carrier flow diagram of the system and the processes. Neglect the acceleration of the fluid. For the fluid, assume a density of 800 kg/m³, and a viscosity of 0.20 Pa·s.

SOLUTION :

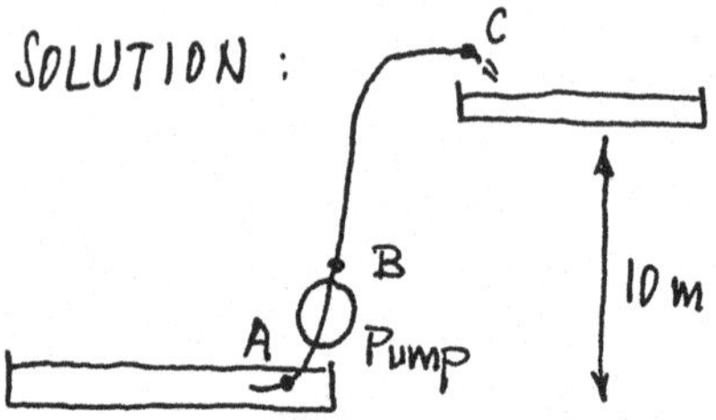

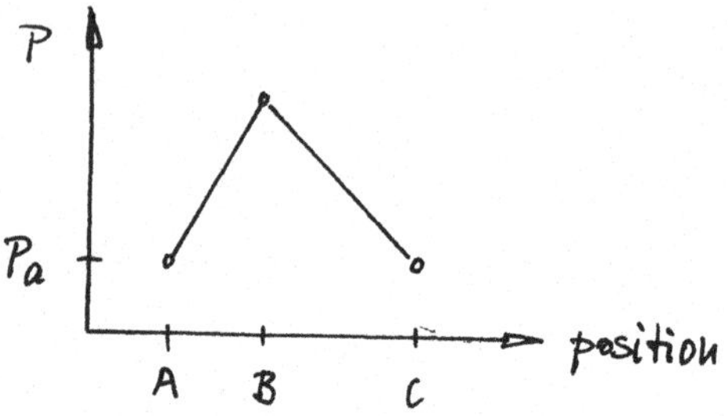

The decrease of the pressure from B to C is the result of the combination of increase in height and fluid resistance because of flow through the pipe:

$$\Delta P_{B\to C} = -\rho g|\Delta h| - \frac{8\eta l}{\pi r^4}|I_V| \quad ; \quad I_V = \frac{1}{\rho} I_m$$

$$= -800 \cdot 9.81 \cdot 10 \text{ Pa} - \frac{8 \cdot 0.20 \cdot 10}{\pi (0.025)^4} \frac{10}{800} \text{ Pa}$$

$$= -7.85 \cdot 10^4 \text{ Pa} - 1.63 \cdot 10^5 \text{ Pa} = -2.42 \cdot 10^5 \text{ Pa}$$

$$|\mathcal{P}_{pump}| = |\Delta P_{pump}||I_V| = |\Delta P_{A\to B} \cdot I_V|$$

$$= |\Delta P_{B\to C} \cdot I_V|$$

$$= 2.42 \cdot 10^5 \text{ Pa} \cdot \frac{10}{800} \text{ m}^3/\text{s} = 3.02 \text{ kW}$$

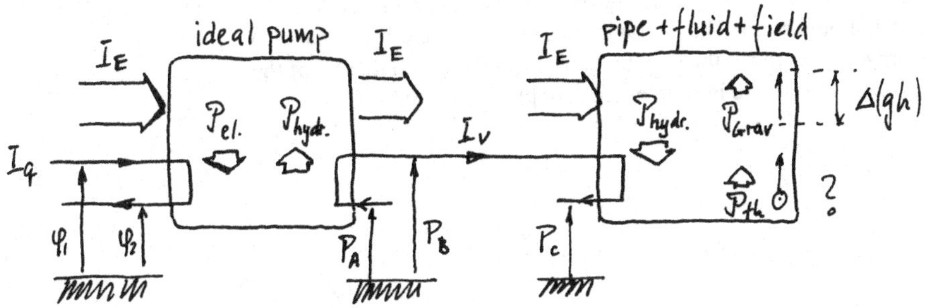

Energy is supplied to the pump and released in an electrical process ($\mathcal{P}_{el.}$). The energy which is released is used ($\mathcal{P}_{hydr.}$) to increase the pressure of the fluid as it is flowing through the pump. On its way from B to C, the energy taken up by the fluid is released again ($\mathcal{P}_{hydr.}$ in the second system), causing two parallel processes: the fluid is lifted in the gravitational field ($\mathcal{P}_{Grav.}$, the energy used in this process is stored in the field), and heat is produced because of friction. This latter process – and what happens to the energy involved ($\mathcal{P}_{th}$) – will be explained in Chapter 1.

■

A tank is filled through a pipe at the bottom (as in the previous problem). Assume the flow through the pipe and in the container not to be affected by friction. a) Express the energy needed to fill the tank up to a certain height in therms of the hydraulic capacitance. b) Where has the energy supplied gone to?

SOLUTION :

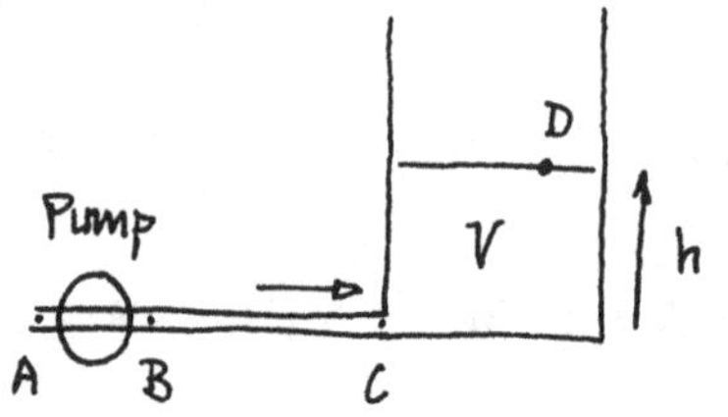

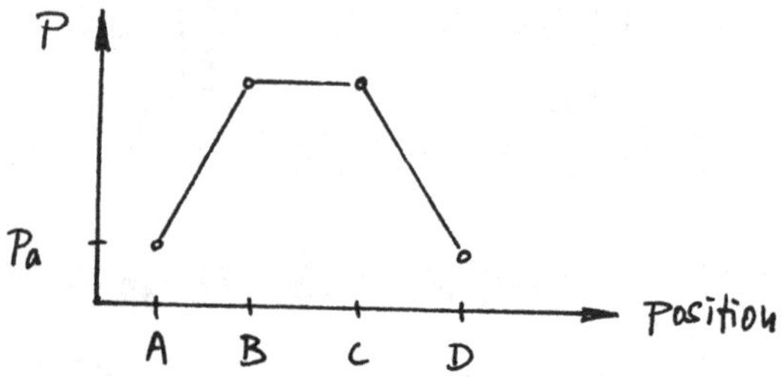

$$P_B = P_a + \rho g h$$

a)

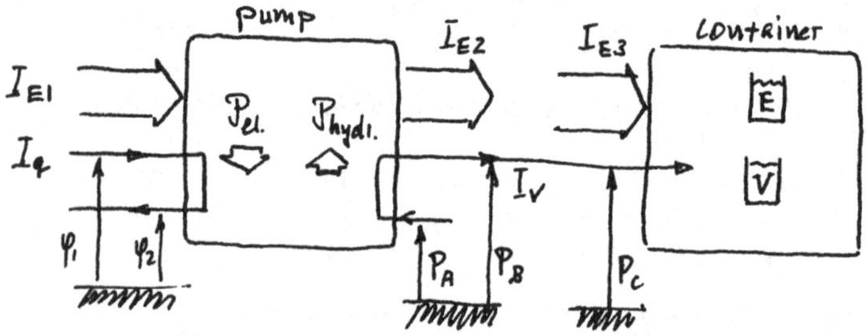

The energy supplied to the pump (I_{E1}) is released ($\mathcal{P}_{el.}$) and then used in the hydraulic process ($\mathcal{P}_{hydr.}$). The fluid transfers the energy to the container (I_{E2}, I_{E3}) where both the fluid and the energy are stored.

Balance of energy : $\Delta E = \int I_{E3}\, dt \; ; \; |I_{E3}| = |I_{E2}| = |\mathcal{P}_{hydr.}| = (P_B - P_A)|I_V|$

Balance of volume: $\dot{V} = -I_V \quad \rightarrow \quad \Delta E = \int_0^t (P_B - P_A)\dot{V}\, dt$

$$\rightarrow \quad \Delta E = \int_0^{P_f} (P_B - P_A)\, dV = \int_0^{h_f} \rho g h A\, dh$$

$$= \rho g A \frac{1}{2} h_f^2 = \frac{1}{2} \frac{\rho g}{A} V_f^2$$

$$= \frac{1}{2} \frac{1}{K} V_f^2$$ where K is the capacitance of the container.

b) As mentioned before, the energy must now be stored in the system. The result just computed is the same as the one known from electricity (the charged capacitor). Here is a simple (hydraulic) image for how much energy is stored together with the volume stored:

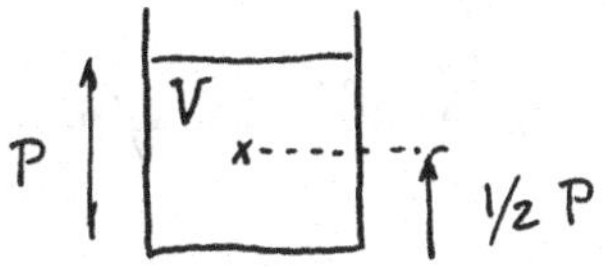

$$E = \underbrace{\frac{1}{2} P}_{\text{average "height"}} \cdot \underbrace{V}_{\text{content}} = \frac{1}{2} K P^2$$

The image can be transferred to other fields of physics and to "containers" which do not necesserily have straight walls (variable capacitance!). ■

Two tanks (see Figure 26) contain oil with a density of 800 kg/m^3 and a viscosity of 0.20 Pa·s. Initially, in the container, which has a cross section of 0.010 m^2, the fluid stands at a level of 10 cm; in the second container (cross section 0.0025 m^2) the level is 60 cm. The hose connecting the tanks has a length of 1.0 m and a diameter of 1.0 cm.

a) What is the volume current right after the hose has been opened? b) Calculate the pressure at A, B, C, and D at this point in time. The pressure of the air is equal to 1.0 bar, and C is in the middle of the hose. c) Determine the rate at which energy is released at the beginning as a consequence of fluid friction. d) Sketch the levels in the containers as a function of time. e) Sketch an electric circuit which is equivalent to the system of containers and pipe. f) Sketch a pressure profile (pressure as a function of position) for a path leading from A to D; include a point C* at the other end of the pipe from point B.

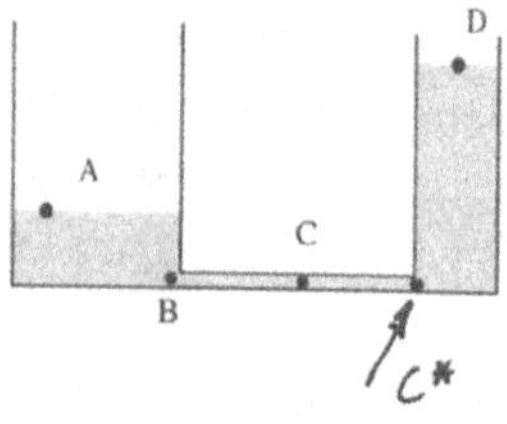

FIGURE 26. Problem 12.

SOLUTION:

a) Resistance law for pipe: $|I_V| = |\Delta P_R| / R_V$

$$|\Delta P_R| = |\Delta P_{B \to C^*}| = \rho g (h_2 - h_1) = 800 \cdot 9.81 (0.60 - 0.10)\ \text{Pa} = 3.92 \cdot 10^3\ \text{Pa}$$

$$R_V = \frac{8 \eta l}{\pi r^4} = \frac{8 \cdot 0.20 \cdot 1.0}{\pi (0.005)^2} \frac{\text{Pa} \cdot \text{s}}{\text{m}^3} = 8.15 \cdot 10^8\ \text{Pa} \cdot \text{s/m}^3$$

$$\rightarrow |I_V| = 3.92 \cdot 10^3 / 8.15 \cdot 10^8\ \text{m}^3/\text{s} = 4.81 \cdot 10^{-6}\ \text{m}^3/\text{s}$$

b) $P_A = P_a = 1.0$ bar ; $P_D = P_a = 1.0$ bar

$P_B = P_A + \rho g h_1 = 1.0078$ bar ; $P_C = P_B + \frac{1}{2} R_V |I_V| = 1.027$ bar

The fluid flows from C* to B, therefore $P_C > P_B$.

c) The hydraulic process in the pipe leads to the release of energy: $|\mathcal{P}_{hydr.}| = |\Delta P_R \cdot I_V| = 1.89 \cdot 10^{-2}$ W

d) The fluid levels in the containers tend toward their final common value:

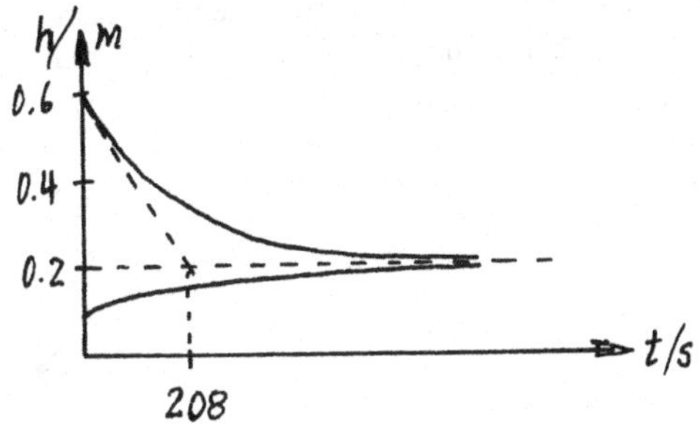

$$A_1 h_{1i} + A_2 h_{2i} = h_f (A_1 + A_2)$$

$$\Rightarrow \quad h_f = \frac{A_1 h_{1i} + A_2 h_{2i}}{A_1 + A_2} = 0.20 \text{ m}$$

The levels grow or decay exponentially. If the small container discharged at a constant rate equal to the value at $t = 0$, it would take

$$\Delta t = 0.0025\,(0.60 - 0.20)/4.81 \cdot 10^{-6} \text{ s} = $$
$$= 208 \text{ s}$$

for the level to drop from 0.60 m to the final value of 0.20 m. It turns out that this value is equal to the hydraulic time constant τ calculated according to

$$\tau = R_V K_V \quad \text{with} \quad K_V = (1/K_1 + 1/K_2)^{-1}$$

e)

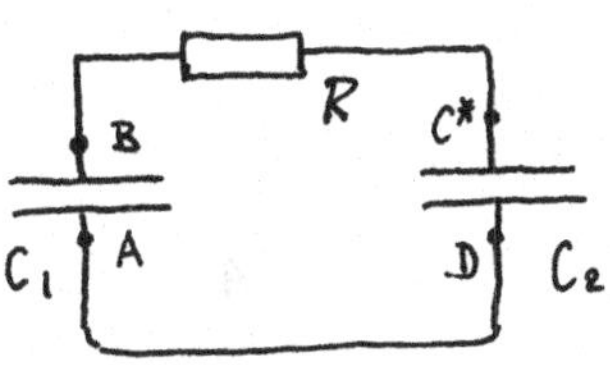

f)

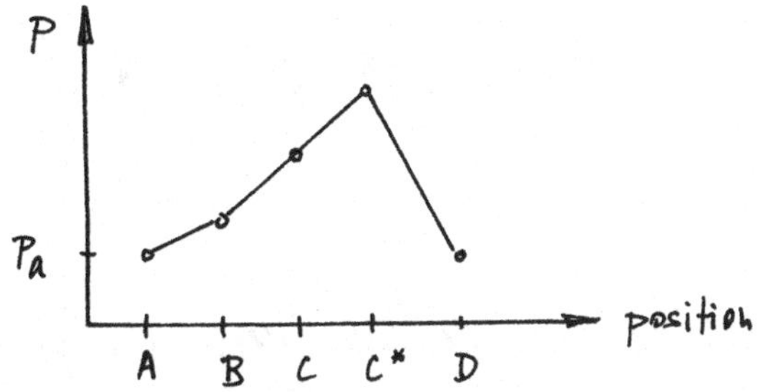

What is the hydraulic inductance for a pipe which is channeling water down from an artificial lake to a hydraulic power plant (Figure 27)? The diameter of the pipe is 0.40 m. How large a rise in pressure would you expect if the volume flux (which has a magnitude of 1.0 m³/s) is reduced to zero within one second?

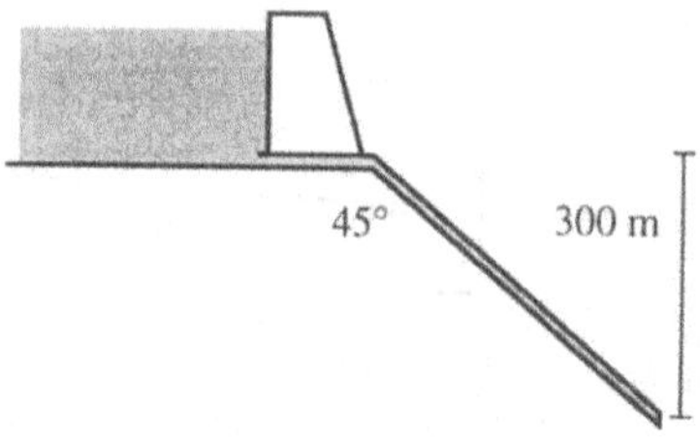

FIGURE 27. Problem 13.

SOLUTION:

Hydraulic inductance (Equation 15):

$$L_V = l\rho/A = \frac{300}{\cos 45^\circ} 1000 \frac{1}{\pi (0.20)^2} \frac{kg}{m^4}$$

$$= 3.38 \cdot 10^6 \ kg/m^4$$

Law of induction:

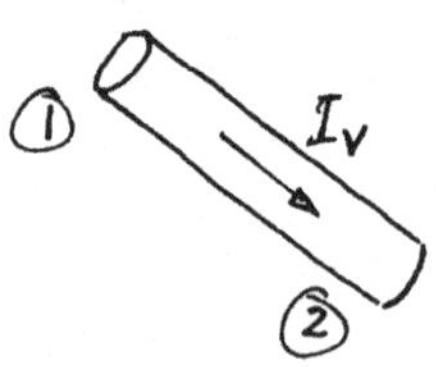

$$\Delta P_L = -L \frac{dI_V}{dt}$$

$$= -3.38 \cdot 10^6 \ kg/m^4 \ \frac{(0-1.0)\ m^3/s}{1.0\ s} = 3.38 \cdot 10^6 \ Pa$$

$$= 33.8 \ bar$$

Because of the inductive effect, the pressure of the water will be larger at point 2 than at point 1.

■

Consider two capacitors, one of them charged, connected in a circuit.
a) Calculate the final charges and voltages of the capacitors in terms of the initial charge and the capacitances. b) Is the energy of the capacitors conserved? c) Translate the problem into an equivalent hydraulic one.

SOLUTION:

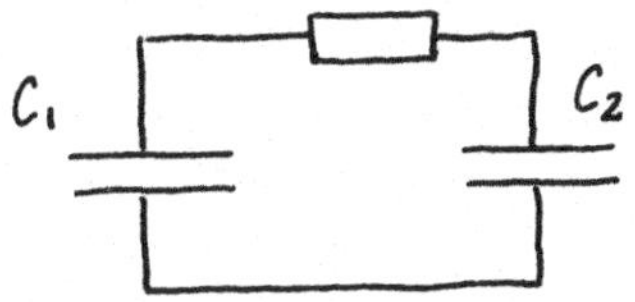

a) Balance of charge:

$$q_{1i} + 0 = q_{1f} + q_{2f}$$

Voltages: $|U_{1f}| = |U_{2f}| = |U_f|$

i: initial, f: final

Balance of charge: $C_1 U_{1i} = C_1 U_{1f} + C_2 U_{2f} = (C_1 + C_2) U_f$

$$\Rightarrow \quad U_f = \frac{C_1}{C_1 + C_2} U_{1i}$$

$$q_{1f} = \frac{C_1^2}{C_1 + C_2} U_{1i} \quad , \quad q_{2f} = \frac{C_1 C_2}{C_1 + C_2} U_{1i}$$

b) $E_i = \frac{1}{2} C_1 U_{1i}^2$

$$E_f = \frac{1}{2}(C_1 + C_2) U_f^2 = \frac{1}{2} \frac{(C_1 + C_2) C_1^2}{(C_1 + C_2)^2} U_{1i}^2 = \frac{1}{2} \frac{C_1^2}{C_1 + C_2} U_{1i}^2 < E_i$$

c)

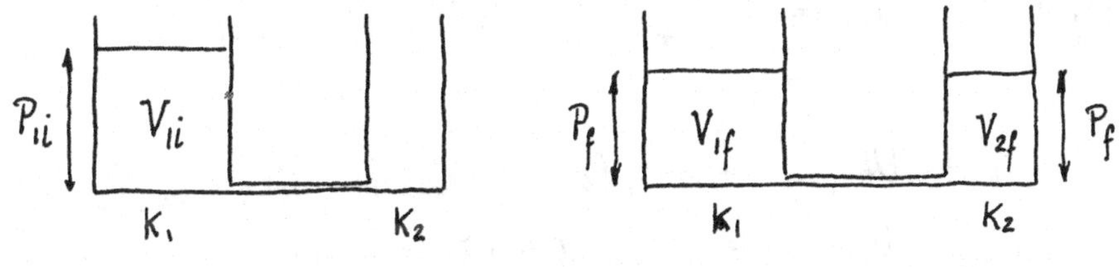

A capacitor (capacitance 150 μF) and a resistor (resistance 1500 Ω) are connected in series to a battery (voltage 50 V) at time $t = 0$ s. The initial charge of the capacitor is equal to zero. a) Derive the equation of balance of the charge of the capacitor. From its solution derive the formula for the electric current as a function of time. b) Draw the carrier and energy flow diagrams for the battery, the resistor, and the capacitor. c) What are the values of the electrical power of the three elements after 0.15 s? d) What are the values of the corresponding electrical energy currents at that point in time? e) Calculate the rate of change of the energy of the capacitor. f) How large is the rate of change of the energy of the resistor?

SOLUTION:

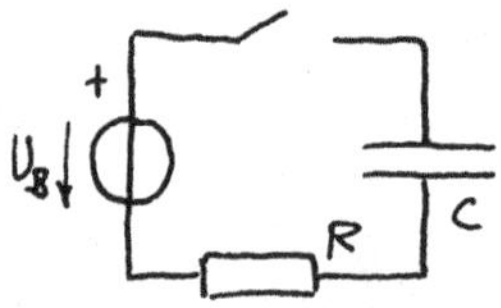

a)

$$\dot q = -I_q$$

$$\dot q = \dot U_C \cdot C$$

$$I_q = -U_R / R$$

$$U_B + U_C + U_R = 0 \quad \text{with } U_B < 0$$

$$\rightarrow \quad U_C(t) = |U_B|\left(1 - e^{-t/\tau}\right) \qquad \tau = R \cdot C$$

$$\rightarrow \quad U_R(t) = |U_B|\, e^{-t/\tau} \quad \rightarrow \quad I_q(t) = -\frac{|U_B|}{R}\, e^{-t/\tau}$$

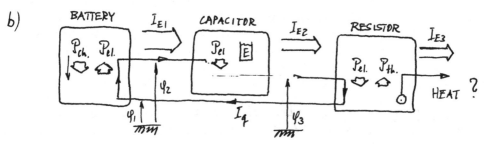

b)

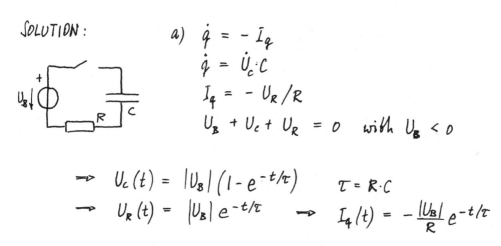

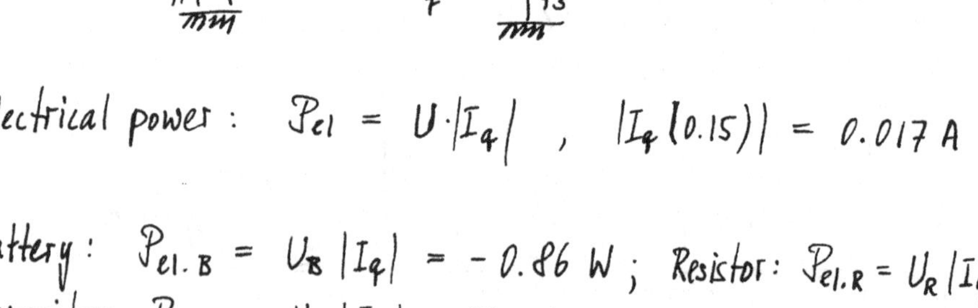

c) Electrical power: $P_{el} = U \cdot |I_q|$, $|I_q(0.15)| = 0.017$ A

Battery: $P_{el.B} = U_B\,|I_q| = -0.86$ W ; Resistor: $P_{el.R} = U_R\,|I_q| = 0.44$ W

Capacitor: $P_{el.C} = U_C\,|I_q| = 24.3\text{ V} \cdot 0.017\text{ A} = 0.42$ W

d) I_{E1} and I_{E2} are energy currents associated with electrical currents. They are calculated by $I_E = \varphi \cdot I_q$. Let us set $\varphi_1 = 0\,V$, which means that $\varphi_2 = 50\,V$ (constant), and $\varphi_3 = 25.7\,V$ (at $t = 0.15\,s$).

$$\rightarrow \quad |I_{E1}| = 50\,V \cdot 0.017\,A = 0.86\,W$$

$$|I_{E2}| = 25.7\,V \cdot 0.017\,A = 0.44\,W$$

e) Balance of energy for capacitor:

$$\dot{E}_C = -(I_{E1} + I_{E2}) = -(-0.86\,W + 0.44\,W) = 0.42\,W$$

f) Balance of energy for resistor:

$$\dot{E}_R = -(I_{E2} + I_{E3})$$

Since we do not know precisely what happens to the energy associated with the process of production of heat, we cannot be precise about our answer. In general, if the resistor heats up during the process, some of the energy communicated to it will be stored there (temporarily).

■

Derive the equation for the oscillation of water in a U-tube (this can be done by combining the equation of balance of volume and the expression for the current in the presence of inductance and resistance). What is the form of the equation if you neglect the effects of resistance? Demonstrate that the period of oscillation can be calculated from

$$\omega = \frac{1}{\sqrt{L_V K_V}}$$

SOLUTION:

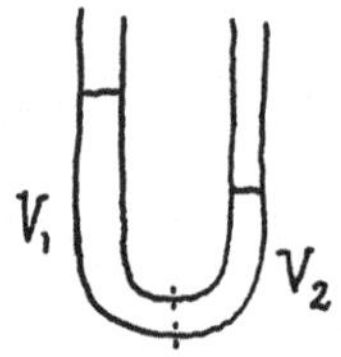

Laws of balance: $\dot{V}_1 = -I_{V1}$

$\dot{V}_2 = -I_{V2}$

"Interaction": $I_{V1} = -I_{V2}$

Capacitance laws: $K_1 \dot{P}_1 = \dot{V}_1$

$K_2 \dot{P}_2 = \dot{V}_2$

where $K_1 = K_2$ are the capacitances of the sides of the U-tube.

Law of induction: $\Delta P_L = -L_V \dot{I}_V$

We have $\Delta P_L = \Delta P = P_2 - P_1$, since there is no fluid resistance. We now calculate the difference of the pressures and take the derivative in preparation of using the law of induction:

$$K_1 (\dot{P}_2 - \dot{P}_1) = 2 I_V \quad \Rightarrow \quad K_1 (\Delta P)^{\cdot} = 2 I_V$$

$$K_1 (\Delta P)^{\cdot\cdot} = 2 \dot{I}_V = -2 \Delta P / L_V$$

$$\Rightarrow \quad (\Delta P)^{\cdot\cdot} + \frac{1}{K_V L_V} \Delta P = 0$$

where $K_V = 1/2 K_1$ is the capacitance of the two sides of the U-tube combined.

$$\rightarrow \quad \omega = 1/\sqrt{K_V L_V}$$

■

A person is pulling a crate across the floor at constant speed via a rope (Figure 30). Take the positive direction to coincide with the direction in which the body is pulled. a) Identify the closed circuit through which the horizontal component of momentum is flowing. b) Determine the momentum fluxes (and their signs) with respect to the crate, the person, and the earth, i.e., the floor. c) There exist several relationships between the different fluxes identified in (b). Which of these have to do with the action-reaction principle (Newton's third law)? Which condition is expressed by the other relationships?

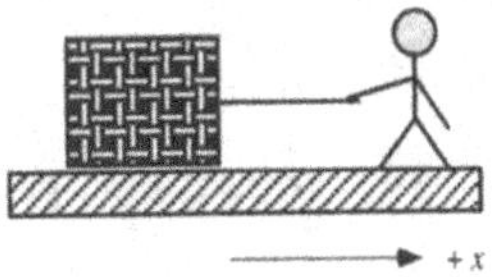

FIGURE 30. Problem 25.

SOLUTION:

a)

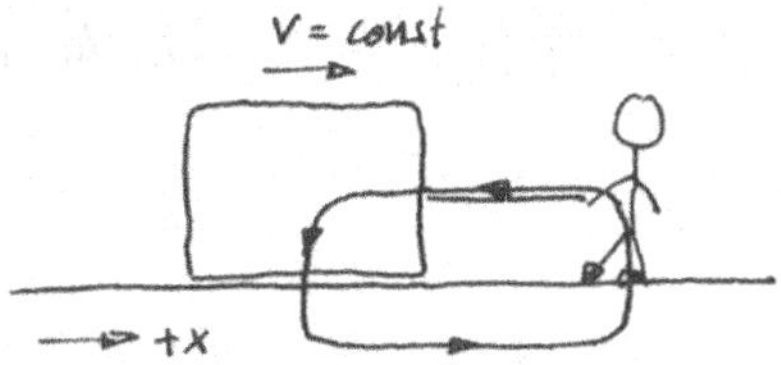

Momentum is supplied to the crate by the person pulling. Therefore, momentum flows through the rope under tension in the negative direction.

Since the momentum of the crate does not change, it must communicate the momentum it receives to the floor. On the other hand, the person picks up the momentum transferred to the crate from the floor. There must be a closed circuit for momentum transport.

b)

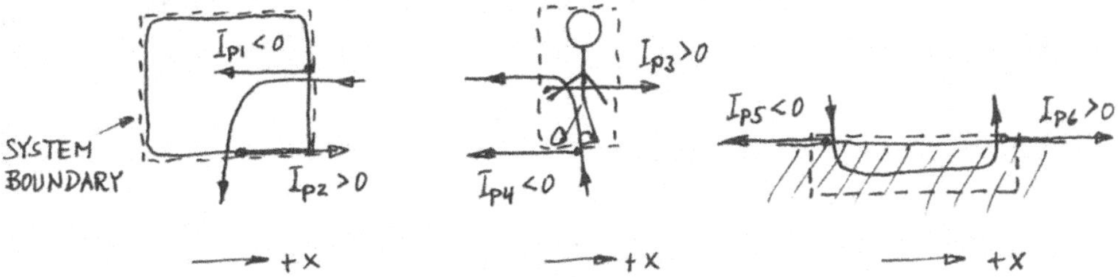

Remember that the flux of a current flowing into a system is counted as a negative number. Instead of the momentum fluxes, you could have drawn the forces $F = -I_p$.

c) For example, $I_{p2} = - I_{p5}$ (momentum leaving the crate enters the floor), or $I_{p1} = - I_{p2}$ (momentum entering the crate must leave again).

The former relation is an instance of the action-reaction principle (which always holds), whereas the latter is a special case of Newton's second law $\dot{p}_{crate} = -(I_{p1} + I_{p2})$ with $\dot{p}_{crate} = 0$ (steady state).

Note: If we had chosen the opposite convention for the signs of fluxes, the law of balance of momentum would be

$$\dot{p} = I_{p1} + I_{p2}$$

and we would have $F = I_p$. We could then have drawn the momentum flux arrows in the previous pictures just like force vectors. Our convention is the one normally used in continuum physics.

■

A body with a mass of 2.0 kg is hanging from a rope. Take the direction downward as the positive one. Determine the fluxes and the sources of the vertical component of momentum with respect to the body. Sketch the flow of momentum through body and rope. How would you introduce forces in this case? What is their relationship to the fluxes and sources expressed as vector quantities?

SOLUTION:

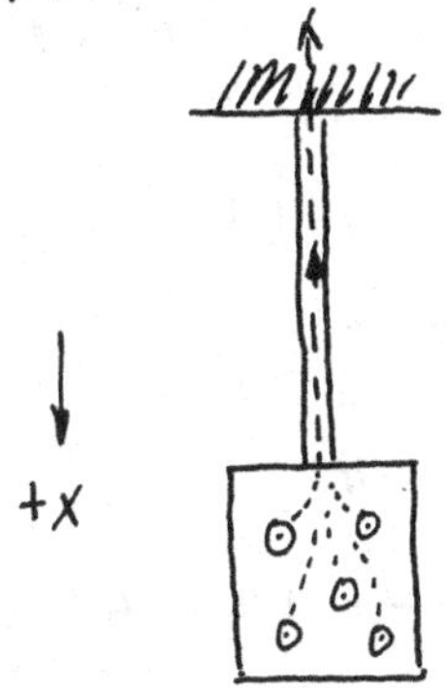

With the positive direction downward, the body receives momentum from the field (⊙: sources of momentum). Since the body does not accelerate, its momentum remains equal to zero. Therefore, the momentum supplied must leave the body through the rope. The rope is under tension (flow of momentum in the negative direction).

Free body diagram

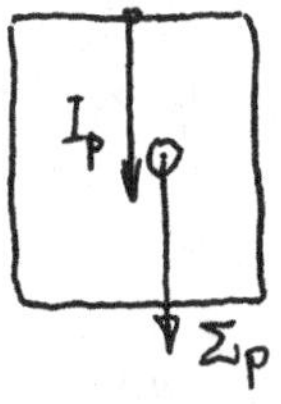

Free body diagram with forces:

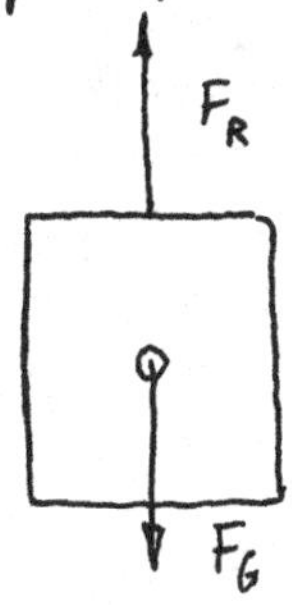

Relationships:

$$F_G = \Sigma_p$$
$$F_R = -I_p$$

■

A rope with a given (constant) mass per length is hanging from a hook. Express the equation of balance of momentum for small segments of the rope and derive the appropriate differential equation for the continuous case. Then determine the momentum current density in the rope as a function of position. How does this quantity relate to the mechanical stress in the rope?

SOLUTION:

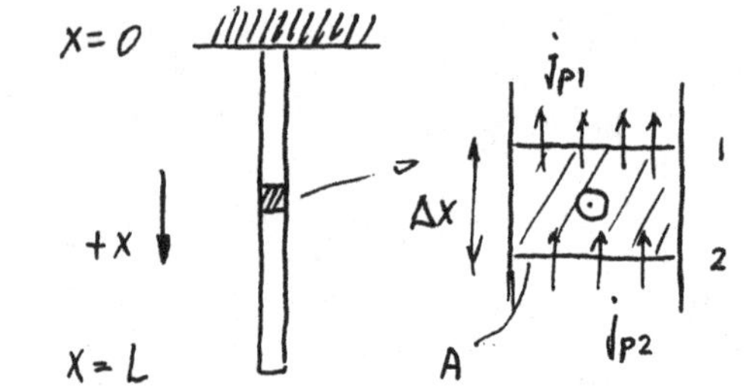

Balance of momentum for segment:

$$\dot{p} = -\left(I_{p1} + I_{p2}\right) + \Sigma_p$$

$$0 = j_{p1} A - j_{p2} A + mg$$

$$= -\left(j_{p2} - j_{p1}\right) A + A \Delta x \rho g$$

$$\Rightarrow \quad \frac{dj_p}{dx} = \rho g$$

Solution of DE: $\quad j_p(x) = \rho g x + j_p(0)$

Boundary value: $\quad j_p(L) = 0$

$$\Rightarrow \quad 0 = \rho g L + j_p(0) \quad \Rightarrow \quad j_p(0) = -\rho g L$$

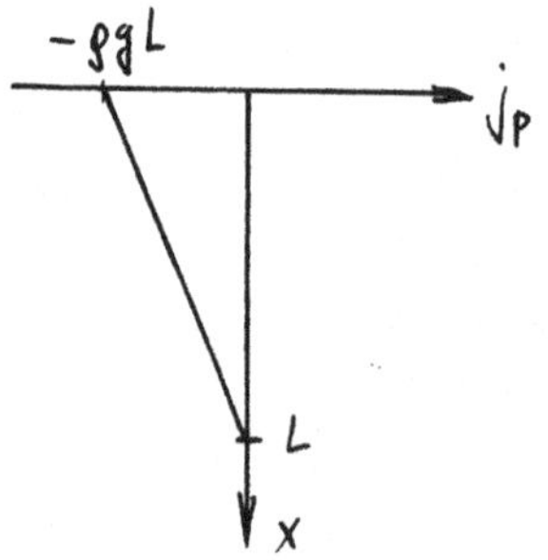

$$\Rightarrow \quad j_p(x) = -\rho g L + \rho g x$$

The fact that $j_p(x)$ is negative means that momentum flows in the negative direction. The rope is under tension.

■

A liquid having a density of 920 kg/m³ is flowing through a pipe whose diameter decreases from 3.0 cm to 1.5 cm (Figure 31). The speed of flow at the smaller exit is 4.82 m/s. a) Compute the convective momentum fluxes at the inlet and the outlet. b) The pressure of the fluid at the inlet is 1.10 bar. Calculate the conductive momentum flux at the entrance. Compare the magnitude of the convective and the conductive fluxes.

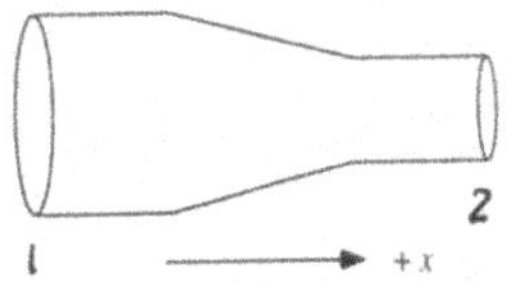

FIGURE 31. Problem 32.

SOLUTION: Preparation: compute speed and pressure at inlet and outlet, respectively:

Balance of volume: $A_1 v_1 = A_2 v_2$

Bernoulli: $P_1 + \frac{1}{2}\rho v_1^2 = P_2 + \frac{1}{2} v_2^2$

$\Rightarrow P_2 = 1.00 \cdot 10^5\,\text{Pa}$, $v_1 = 1.205\,\text{m/s}$

a)

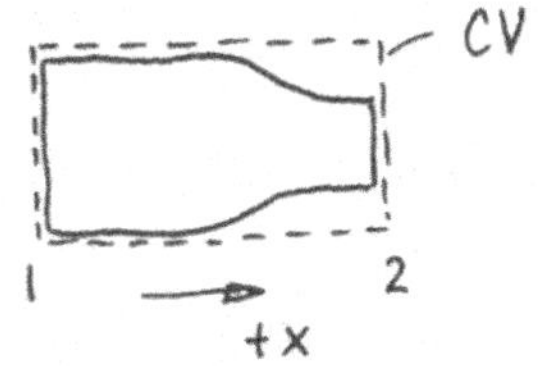

$$I_{p,conv,1} = v_1 I_{m1} = 1.205\,\text{m/s}\left(-1.205 \cdot \pi (0.015)^2 \cdot 920\right)\text{kg/s} = -0.944\,\text{N}$$

$$I_{p,conv,2} = v_2 I_{m2} = 4.82\,\text{m/s} \cdot 4.82 \cdot \pi (0.0075)^2\, 920\,\text{kg/s} = 3.78\,\text{N}$$

b)

$$I_{p,cond,1} = -P_1 A_1 = -1.10 \cdot 10^5\,\text{Pa} \cdot \pi (0.015)^2\,\text{m}^2 = -77.7\,\text{N}$$

$$I_{p,cond,2} = P_2 A_2 = 1.00 \cdot 10^5\,\text{Pa} \cdot \pi (0.0075)^2\,\text{m}^2 = 17.66\,\text{N}$$

The conductive momentum transports (due to pressure) are considerably larger than the convective ones. ■

In the previous problem compute the force which holds the pipe in place. The pressure of the surrounding air is 1.0 bar. Take the flow through the pipe to be ideal.

SOLUTION:

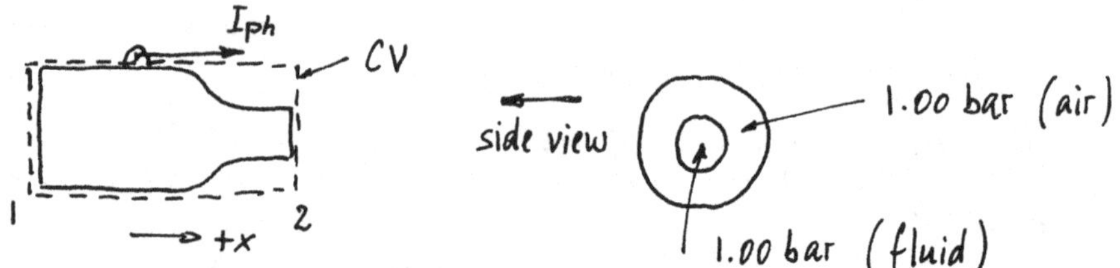

I_{ph} : momentum flux as a result of holding the pipe

Balance of momentum for CV:

$$\dot{p} = -\left(I_{p,cond,1} + I_{p,conv,1} + I_{p,cond,2} + I_{p,conv,2} + I_{ph}\right)$$

$$\dot{p} = 0 \quad \text{(steady-state conditions)}$$

$$0 = -\left(-77.7\,N - 0.944\,N + \underbrace{P_2\,\pi\,(0.015)^2\,N}_{\text{due to pressure over entire cross section of control volume at point 2}} + 3.78\,N + I_{ph}\right)$$

$$\Rightarrow I_{ph} = 77.7\,N + 0.944\,N - 70.65\,N - 3.78\,N = 4.21\,N$$

$$F_h = -I_{ph} = -4.21\,N$$

(The force points in the negative direction.)

■

A rocket is moving through space far from any gravitational fields at constant speed v with respect to an observer. The engines are ignited such that the flux of mass out of the rocket is constant, and the speed of the gases is v_g with respect to the engines. a) Formulate the equation of balance of momentum for the rocket with respect to the observer. b) Formulate the balance of momentum for an observer moving at the same speed as the rocket before the ignition of the engines. c) Solve the equation of motion in both cases.

SOLUTION:

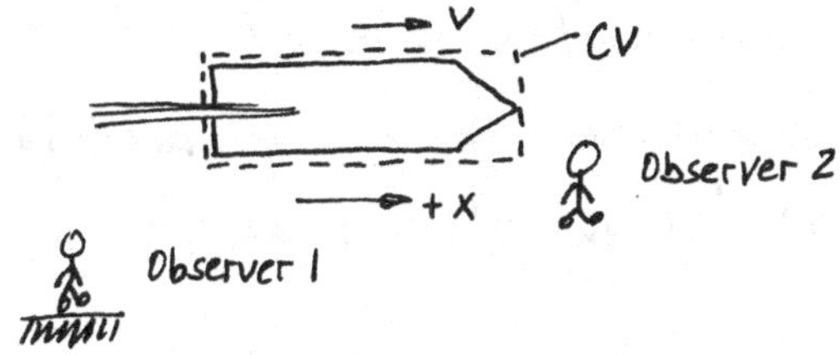

a) $\dot{p} = -I_{p,conv}$

$I_{p,conv} = v_{gas} I_m$ where $v_{gas} = v + v_g$

b) $\dot{p} = -I_{p,conv}$

$I_{p,conv} = v_{gas} I_m$ where $v_{gas} = v_g$

c) Observer 1: $\dot{p} = m\dot{v} + \dot{m}v$, $\dot{m} = -I_m \Rightarrow m\dot{v} + v(-I_m) = -(v+v_g) I_m$

$\Rightarrow m\dot{v} = -v_g I_m$ where $v_g < 0$, $I_m > 0$

Observer 2: $\dot{p} = m\dot{v} + \dot{m}v$, $v = 0 \Rightarrow m\dot{v} = -v_g I_m$ as for Observer 1.

Solution of DE: $m\dfrac{dv}{dt} = v_g \dfrac{dm}{dt} \Rightarrow \int_{v_0}^{v} v\,dv = v_g \int_{m_0}^{m} \dfrac{dm}{m}$

$$\Rightarrow v(m) = v_0 + v_g \ln\left(\frac{m}{m_0}\right) = v_0 + |v_g| \ln\left(\frac{m_0}{m}\right)$$

$$m(t) = m_0 - |I_m|\, t$$

■

An open car moves underneath a vertically falling current of water (Figure 32). At the same time, the car loses water through a hole at the bottom at the same rate at which it picks up water. Set up the equation of motion for the car and determine its speed as a function of time. Assume friction to be negligible.

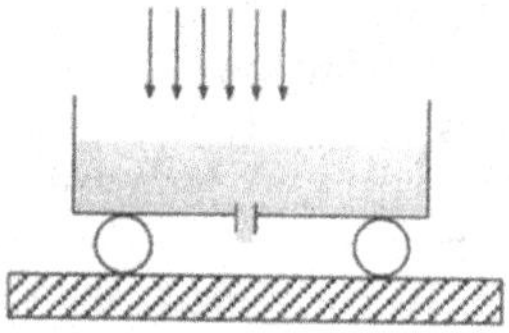

FIGURE 32. Problem 38.

SOLUTION:

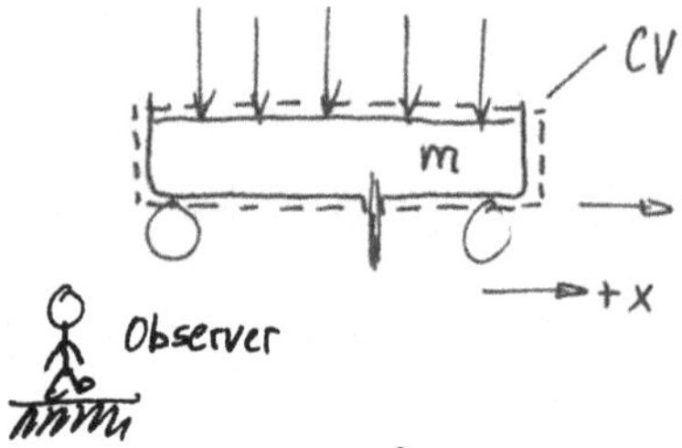

Only the balance of the x-component of momentum is important in the present case. The y-component influences friction, which is absent.

$$\dot{P}_x = -I_{px}$$

I_{px} : x-momentum flux due to water flowing out of the car. The rain does not carry x-momentum into the CV.

$$I_{px} = v I_m \quad ,$$

v : instantaneous speed of car also is the speed of the water

$$\dot{p} = m\dot{v} + v\dot{m} = m\dot{v} \quad \text{since } \dot{m} = 0$$

$$\Rightarrow \quad m\dot{v} = -v I_m \quad , \quad I_m > 0$$

$$\int_{v_0}^{v} \frac{dv}{v} = -\frac{I_m}{m}\int_0^t dt \quad \rightarrow \quad \ln\left(\frac{v}{v_0}\right) = -\frac{I_m}{m}t$$

$$v(t) = v_0 \exp\left(-\frac{I_m}{m}t\right)$$

■

One mole of salt is dissolved in 10 liters of water inside a tank. A current of fresh water of 1.0 kg/s enters the container where it is instantly mixed with the salt-water solution. The solution flows out of the tank at a rate of 0.50 kg/s. a) Determine the concentration of salt inside the tank as a function of time. b) How much salt has left the container in the first 10 s?

SOLUTION:

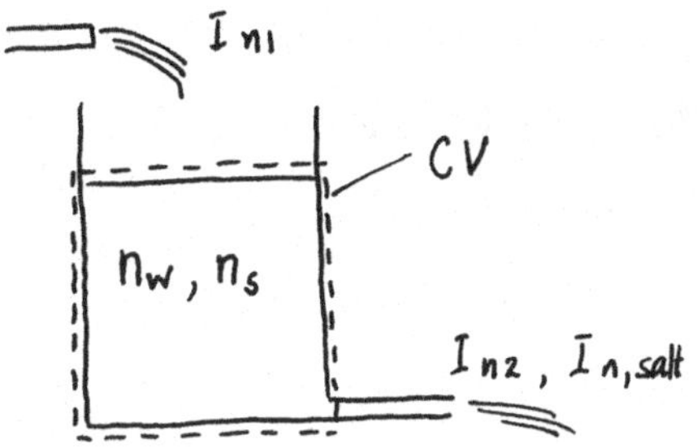

We have to consider the balance of amount of water and amount of salt in the tank:

$$\dot{n}_w = -(I_{n1} + I_{n2})$$

$$\dot{n}_s = -I_{n,salt}$$

$I_{n,salt}$ depends upon the concentration of salt in the water current leaving the system (which is equal to the instantaneous concentration in the control volume), and the water current:

$$I_{n,salt} = c\, I_{n2}$$

where $c = n_s / n_w$

a) Equations of balance: $\dot{n}_w = -\left(-55.5 \text{ mole/s} + 27.8 \text{ mole/s}\right) = 27.8 \text{ mole/s}$

$$\dot{n}_s = -c\, I_{n2} = -\frac{n_s}{n_w} 27.8 \text{ mole/s}$$

$$\rightarrow \quad n_w(t) = 556 \text{ mole} + 27.8 \text{ mole/s} \cdot t$$

$$\frac{dn_s}{dt} = -\frac{27.8 \text{ mole/s}}{556 \text{ mole} + 27.8 \text{ mole/s} \cdot t} \cdot n_s$$

$$\int_{n_{s0}}^{n_s} \frac{dn_s}{n_s} = -b \int_0^t \frac{dt}{a+bt} \qquad a = 556 \text{ mole},\ b = 27.8 \text{ mole/s}$$

$$\Rightarrow \quad \ln\left(\frac{n_s}{n_{s0}}\right) = -\ln\left(\frac{a+bt}{a}\right)$$

$$\Rightarrow \quad n_s(t) = \frac{a}{a+bt}\, n_{s0} \quad , \quad n_{s0} = 1.0 \text{ mole}$$

b) $\Delta n_s = n_{s,e} \quad \Rightarrow \quad n_{s,e} = n_s(10) - n_{s0}$

$$= \left(\frac{556}{556 + 27.8 \cdot 10} - 1\right) 1.0 \text{ mole}$$

$$= -0.33 \text{ mole}$$

■

Two substances A and B are pumped into a reactor at rates of 10 mole/s and 20 mole/s for A and B, respectively. A certain amount of substance B already is present in the reactor. A and B react instantly to form substance C according to A + 2B → 3C. C is pumped out of the reactor at a rate of 15 mole/s. Determine the rates of change of the amounts of A, B, and C in the reactor.

SOLUTION:

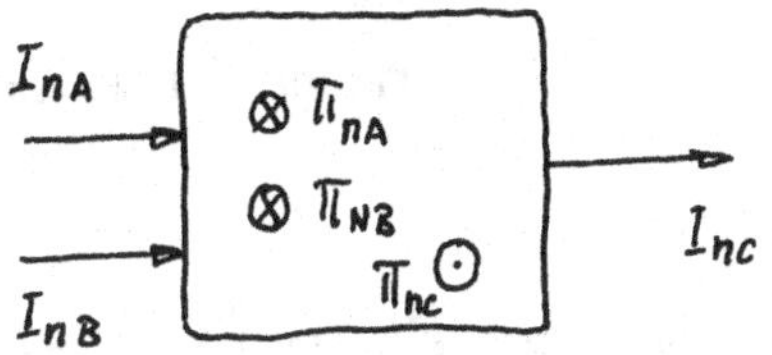

Assume that C is pumped out of the reactor at the same rate as it is produced.

⊗ destruction

⊙ production

Laws of balance:

$$\dot{n}_A = -I_{nA} + \Pi_{nA} \qquad \Pi_{nA} < 0, \quad I_{nA} = -10 \text{ mole/s}$$

$$\dot{n}_B = -I_{nB} + \Pi_{nB} \qquad \Pi_{nB} < 0, \quad I_{nB} = -20 \text{ mole/s}$$

$$\left.\begin{array}{l} \dot{n}_c = -I_{nc} + \Pi_{nc} \\ I_{nc} = \Pi_{nc} \end{array}\right\} \rightarrow \dot{n}_c = 0$$

$$\Pi_{nc} = I_{nc} = 15 \text{ mole/s}$$

$$\rightarrow \left.\begin{array}{l} \Pi_{nA} = -5 \text{ mole/s} \\ \Pi_{nB} = -10 \text{ mole/s} \end{array}\right\} \text{ from the reaction equation}$$

$$\rightarrow \dot{n}_A = 10 \text{ mole/s} - 5 \text{ mole/s} = 5 \text{ mole/s}$$

$$\dot{n}_B = 20 \text{ mole/s} - 10 \text{ mole/s} = 10 \text{ mole/s}$$

■

A fluid is confined between the walls of two concentric cylinders (Figure 33). The gap between the cylinders is very narrow. The inner cylinder can be rotated, and the torque upon it and the angular velocity can be measured. Determine the viscosity of the fluid.

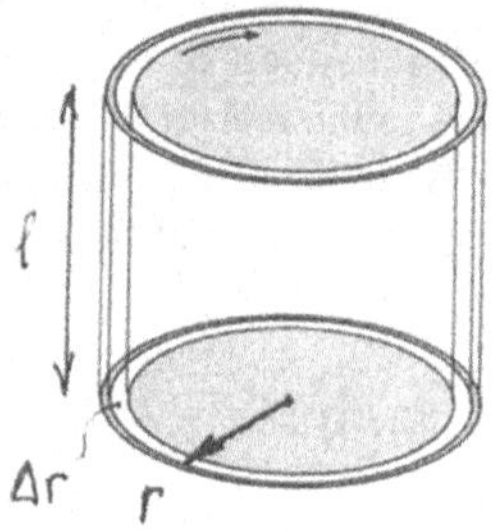

FIGURE 33. Problem 42.

SOLUTION:

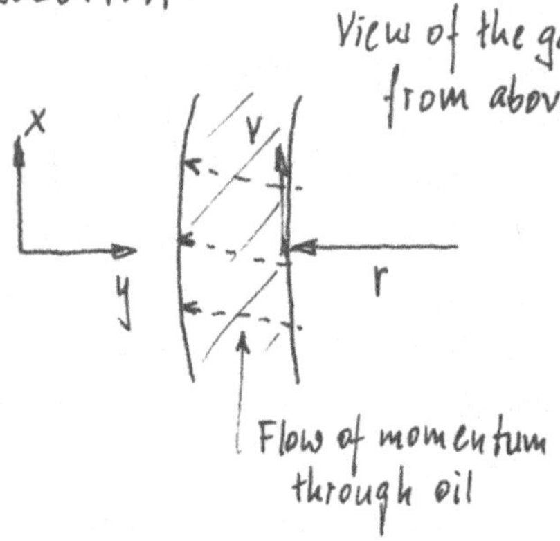

The torque τ and the radius r of the cylinder determine the momentum flux through the mantle of the cylinder:

$$|I_p| = \frac{1}{r}\tau$$

On the other hand: $|I_p| = A\,|j_p| = 2\pi r l\,|j_p|$

The speed of the circumference of the inner cylinder is given by $v = r \cdot \omega$. Now we can determine the viscosity (Equation 35):

$$|j_p| = \mu \left|\frac{dv_x}{dy}\right| = \mu \left|\frac{\Delta v_x}{\Delta r}\right| = \mu \frac{v}{\Delta r}$$

$$\Rightarrow \quad \mu = \frac{\Delta r\,|j_p|}{v} = \frac{\Delta r \dfrac{\tau}{2\pi r^2 l}}{r \cdot \omega} = \frac{\Delta r \cdot \tau}{2\pi r^3 l \omega}$$

∎

Show that if you use momentum as the fundamental quantity, a (linear) spring is an inductor. Determine its inductance. Show that a body hanging from the spring has the property of a capacitor. Now determine the frequency of oscillation from the capacitance and the inductance of the system.

SOLUTION:

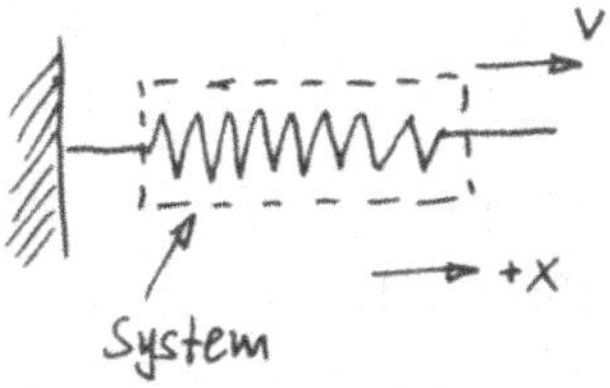

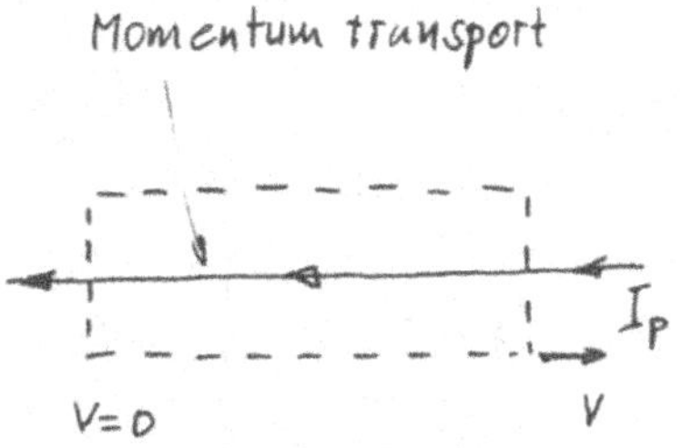

Hook's law: $A \cdot j_p = -D\,\Delta x$ (D: spring constant)

Time derivative: $A \frac{dj_p}{dt} = -D \cdot v$

Interpretation: A change of the position of one end of the spring relative to the other end leads to a change of the momentum current through the spring: $\Delta v \leftrightarrow \dot{I}_p$

$$\Delta v = -\frac{1}{D} A \frac{dj_p}{dt} = -L_p \dot{I}_p \Rightarrow L_p = \frac{1}{D}$$

Since the mass of a body is its momentum capacitance, a body and a spring represent an LC-system:

$$\omega = \frac{1}{\sqrt{C_p L_p}} = \sqrt{\frac{D}{m}}$$

Explain the difference between power of a process and energy currents with respect to a system.

SOLUTION:

Example:

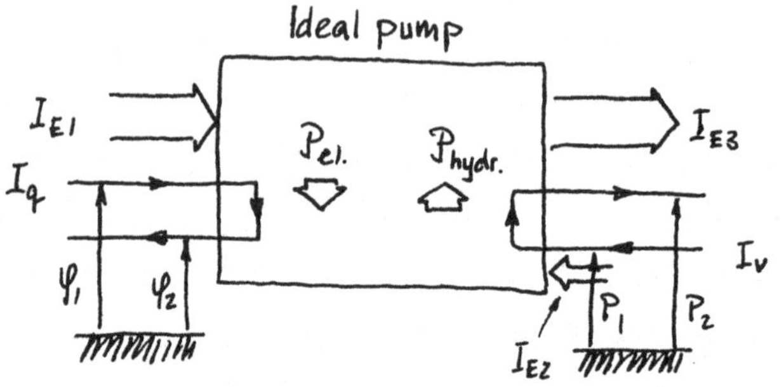

We use the term <u>power</u> for the rate at which energy is released (P_{el}) or used (bound, $P_{hydr.}$) in processes <u>inside</u> a system. Release of energy is the consequence of an "energy carrier" (charge) dropping from higher to lower potential. "Using" energy means binding it to an "energy carrier" which is pumped from low to high potential.

<u>Energy currents</u>, on the other hand, denote the rate of energy transfer to or from a system. Only energy currents appear in laws of balance of energy. ■

Calculate the source rate of energy with respect to a stone falling freely 2.0 s after it has been released. The mass of the body is equal to 0.20 kg. Where does the energy come from?

SOLUTION :

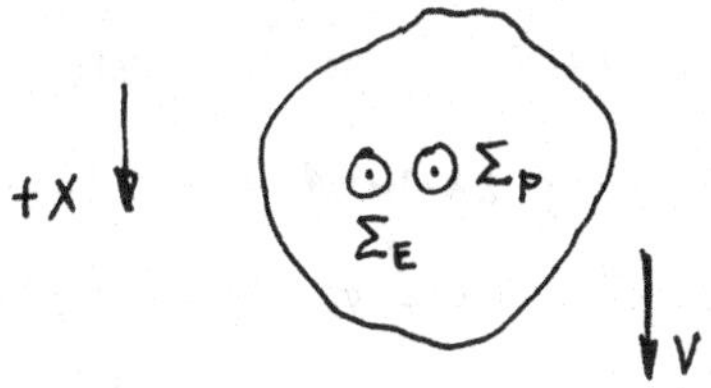

The source rate of energy is calculated from the source rate of momentum and the instantaneous speed of the body :

$$\Sigma_E = v\,\Sigma_p$$

Gravitational sorce rate of momentum : $\Sigma_p = m \cdot g$

Speed after 2.0 s : $v(t) = gt = 9.81 \cdot 2.0 \text{ m/s} = 19.6 \text{ m/s}$

$$\rightarrow \quad \Sigma_E\,(t = 2.0\,s) = 19.6 \cdot 0.20 \cdot 9.81 \text{ W} = 38.5 \text{ W}$$

The energy is said to be supplied from the gravitational field. ■

A car moving horizontally at a constant speed of 120 km/h is using 8.0 liters of gasoline in a distance of 100 km. The mechanical efficiency of the engine is 0.20. Draw a flow diagram for the car as the system, depicting energy carriers, energy currents, and power. How large is the magnitude of the sum of all resistive forces acting upon the body? Repeat the flow diagram with the engine as the system.

SOLUTION:

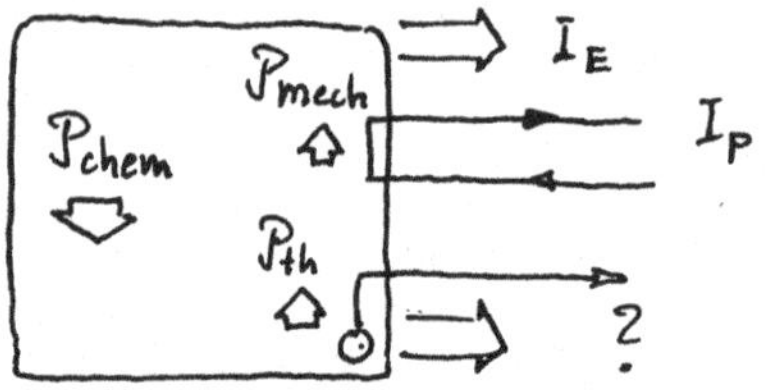

Energy is released by the chemical process of burning fuel. The energy is drawn from the system; it is stored in the fuel. 80% of this energy is used in a thermal process ($\mathcal{P}_{th}$) which is explained in Chapter 1. 20% of the energy released drives the mechanical process which consists of pumping momentum from $v=0$ (the ground) to $v = 120$ km/h. The momentum picked up by the car is flowing out of the system as a result of resistive forces.

Sum of resistive forces: $|I_p|$

$$|\mathcal{P}_{mech}| = |\Delta v \cdot I_p| \quad ; \quad |\mathcal{P}_{mech}| = 0.20\,|\mathcal{P}_{chem}|$$

$$|\mathcal{P}_{chem}| = \frac{8 \cdot 35 \cdot 10^6}{10^5 / 33.3}\ \text{W} = 93.3\ \text{kW}$$

(One liter of gasoline releases about 35 MJ of energy.)

$$\rightarrow |I_p| = \frac{1}{v}\,0.20\,|\mathcal{P}_{chem}| = 560\ \text{N}$$

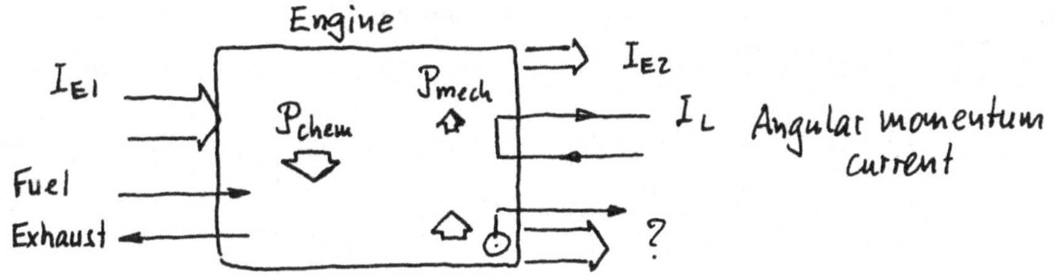

By changing the system boundary, our description commonly must change. Usually, energy currents and carrier currents change. In the present case, energy is supplied to the engine, and energy is removed from the system as a result of a rotational process (apart from the thermal process).

■

A linear spring is attached to a wall on one side. As it is stretched, determine all energy fluxes with respect to the spring. Calculate the change of the energy content of the spring as a function of the stretching.

SOLUTION:

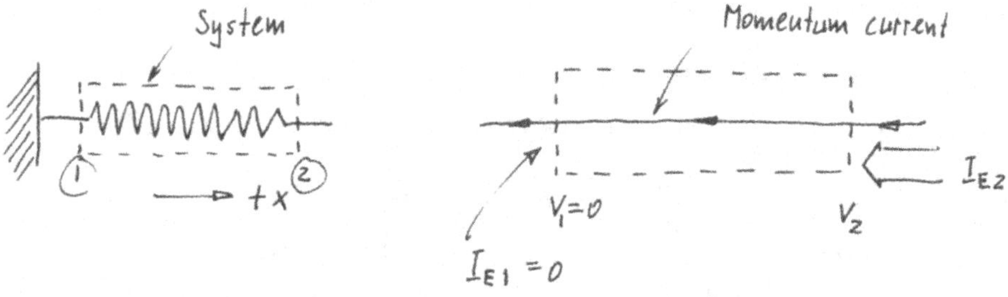

Energy current at point 2 (there is a momentum current, and the boundary is moving):

$$I_{E2} = v_2 I_{p2}$$

$$I_{p2} = -D \cdot x$$ where x is the stretching of the spring

Balance of energy for the spring:

$$\dot{E} = -I_{E2} \quad D v_2 x$$

$$\Delta E = \int_0^t \dot{E}\, dt = \int_0^t D v_2 x \, dt = \int_0^t D x \dot{x} \, dt$$

$$= \int_0^x D x \, dx = \frac{1}{2} D x^2$$

A mill stone grinds wheat, rotating atop a horizontal surface. Draw a flow diagram for energy carriers, energy fluxes, and power for the stone as the system. The stone rotates once in 2.0 s. If the energy flux supplied to the stone is 1.0 kW, how large is the flux of angular momentum through the shaft of the mill stone? How large is the net torque with respect to the stone?

SOLUTION :

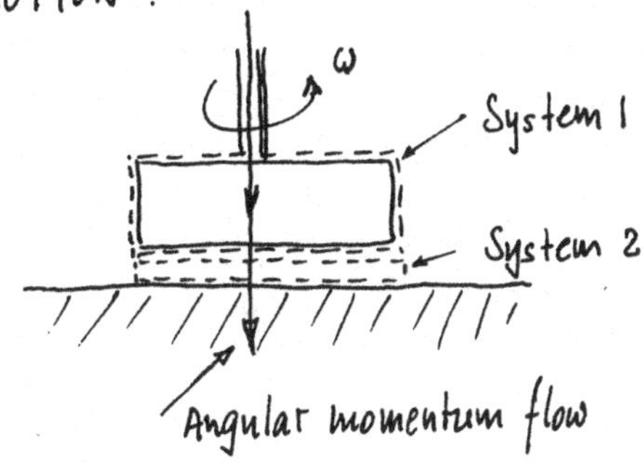

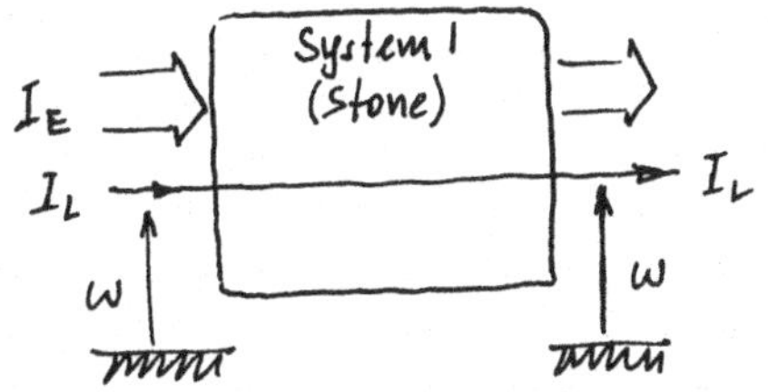

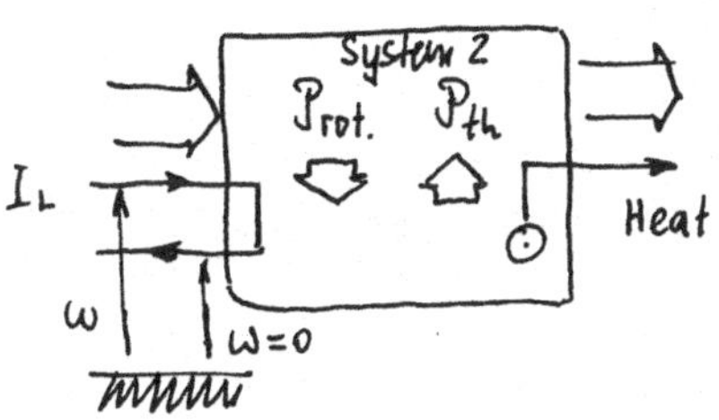

If we treat the stone as a rigid system, and if we only look at the stone, angular momentum simply flows through it at constant "level" ω, from the shaft to the surface. Therefore:

$$|I_{L\,in}| = |I_{L\,out}|$$

and $\dot{L} = 0$. The net torque is equal to zero.

The interesting things actually happen where grinding takes place (System 2). There, angular momentum flows to the surface with $\omega = 0$.

Flux of angular momentum : $|I_E| = |\omega I_L|$

$$\Rightarrow |I_L| = \frac{1}{\omega}|I_E| = \frac{1000}{3.14}\ \text{Nm} = 319\ \text{Nm}$$

In Figure 35 you find a system dynamics diagram of a body suspended from a spring and oscillating up and down. Identify the part of the diagram which represents Newton's law. What type of relation is represented by the box and the flow labeled *position* and *velocity*? What is the nature of the relation represented by the flow called net *momentum flux* and the three variables associated with it? Can you identify the feedback loops of the system?

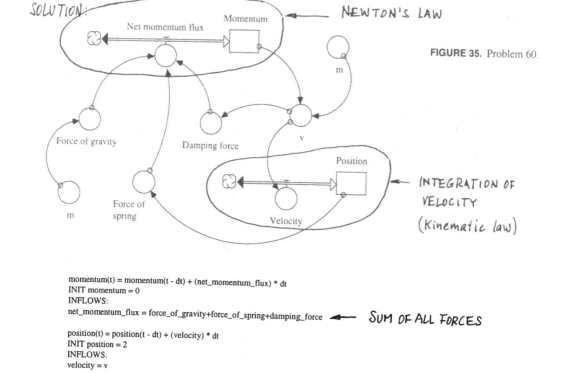

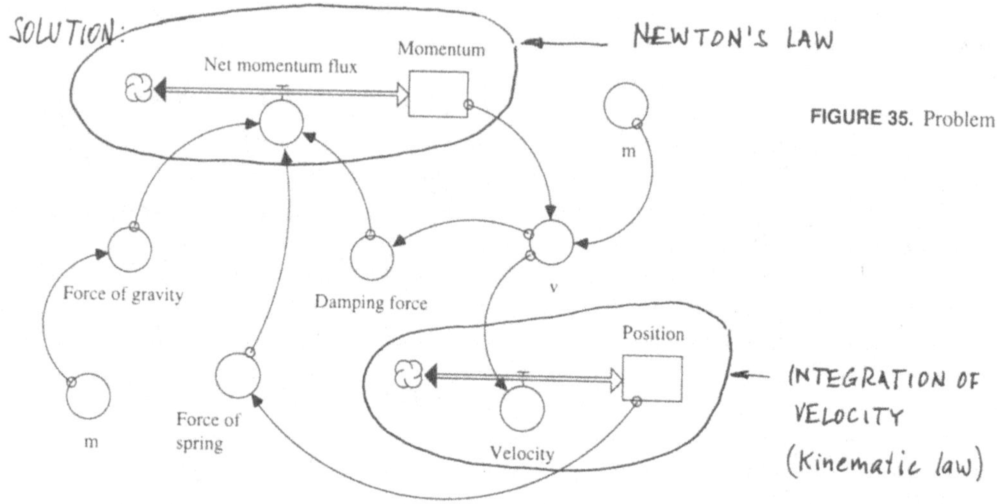

FIGURE 35. Problem 60.

momentum(t) = momentum(t - dt) + (net_momentum_flux) * dt
INIT momentum = 0
INFLOWS:
net_momentum_flux = force_of_gravity+force_of_spring+damping_force ← SUM OF ALL FORCES

position(t) = position(t - dt) + (velocity) * dt
INIT position = 2
INFLOWS:
velocity = v

damping_force = -0.5*v
force_of_gravity = 9.81*m
force_of_spring = -10*position } ← CONSTITUTIVE LAWS FOR FORCES
m = 0.50
v = momentum/m ← CAPACITIVE LAW

There are two feedback loops. One goes from Momentum via v, Damping force, and Net momentum flux back to Momentum. The second one is: Momentum - v - Position - Force of spring - Net momentum flux - momentum.

■

Figure 36 shows the system dynamics model of Problem 39. Identify the graphical representation of the differential equations which you have written for the solution of Problem 39. Does the numerical solution presented correspond to what you have calculated?

SOLUTION:

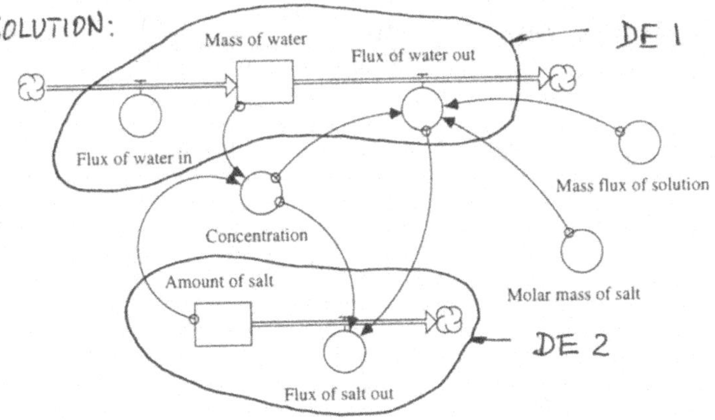

FIGURE 36. Problem 61.

```
amount_of_salt(t) = amount_of_salt(t - dt) + (- flux_of_salt_out) * dt
INIT amount_of_salt = 1
OUTFLOWS:
flux_of_salt_out = concentration*flux_of_water_out
mass_of_water(t) = mass_of_water(t - dt) + (flux_of_water_in - flux_of_water_out) * dt
INIT mass_of_water = 10
INFLOWS:
flux_of_water_in = 1
OUTFLOWS:
flux_of_water_out = mass_flux_of_solution/(1+molar_mass_of_salt*concentration)
concentration = amount_of_salt/mass_of_water
mass_flux_of_solution = 0.50
molar_mass_of_salt = 0.040
```

DE 1:

$$\dot{m}_w = -(I_{m1} + I_{m2})$$

DE 2:

$$\dot{n}_s = - I_{n,salt}$$

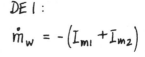

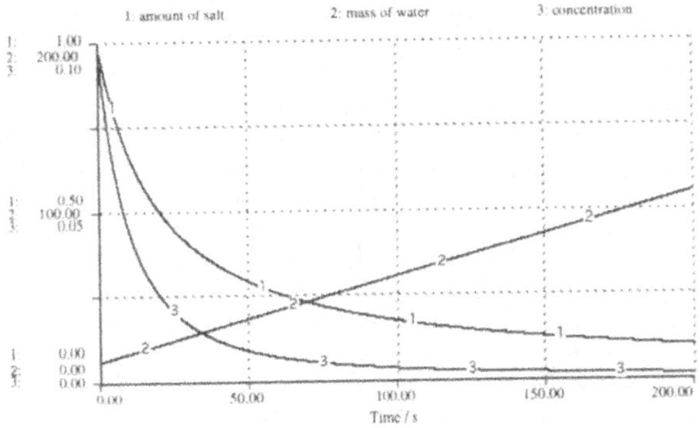

Here, the problem is presented in terms of mass rather than amount of substance for water. Otherwise, things are the same.

As calculated in Problem 39 b, 33% of the initial amount of salt leaves the system in the first 10 s. ■

Sketch a system dynamics model for the process of discharging of a capacitor. Repeat the problem for charging with the help of a battery.

SOLUTION: We will treat the second case (charging of capacitor).

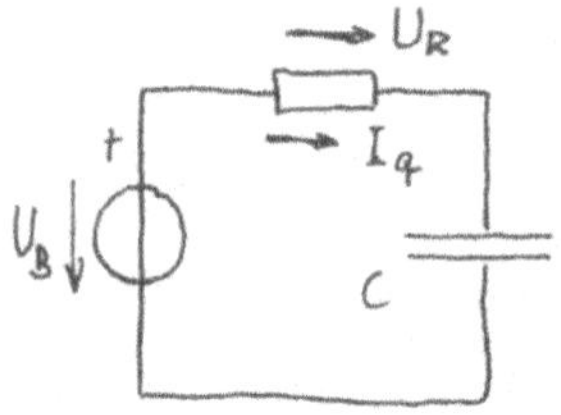

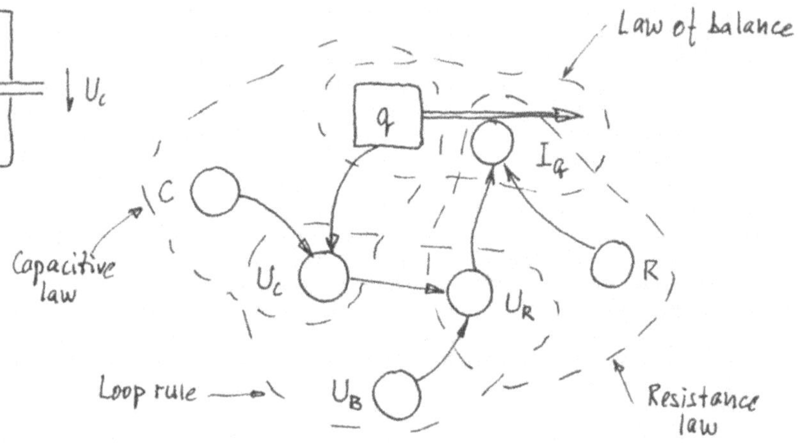

Listing of relations:

Balance of charge of capacitor	$q(t) = q(t-\Delta t) + (-I_q)\Delta t$
Initial value of charge	$q_{init} = 0$
Capacitive law	$U_C = q/C$
Loop rule	$U_R = -U_B - U_C$
Resistance law	$I_q = U_R/R$
Parameters	$C = \ldots\ldots$
	$R = \ldots\ldots$
	$U_B = -12$ (U_B is negative)

■

CHAPTER 1

Solutions of Selected Problems

Consider a moving body that splits into two halves which continue moving along together. Which mechanical quantity is divided among the bodies? Which other mechanical variable is not divided up, leaving each of the parts with its initial value? Compare electrical and thermal phenomena to this mechanical example. Which electrical or thermal quantities correspond to the mechanical variables?

SOLUTION: This problem discusses the issue of extensive versus intensive quantities. A moving body posesses momentum, and it has a certain velocity. [It also has a certain mass; the mass of a body, however, cannot be used as a measure of the state of its motion.] If you divide the body into two parts, you will see that while the velocity of each of the parts is the same as that of the total body, the momentum is divided up among its parts:

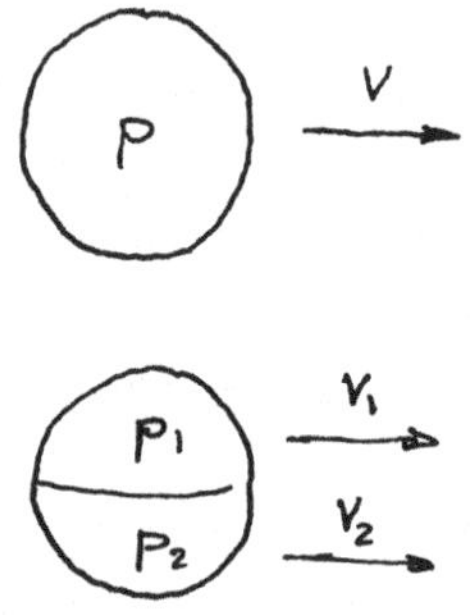

$$v_1 = v_2 = v$$

$$p_1 + p_2 = p$$

Therefore, momentum is said to be the extensive (additive) mechanical quantity (it scales with the 'size' of the body), while the velocity is called the intensive quantity.

Comparison:

	extensive	intensive
motion	momentum	velocity
electricity	charge	el. potential
heat	heat (= entropy)	temperature

■

Why shouldn't we think of energy as a mechanical, electrical or thermal quantity? Why would it be particularly wrong to identify stored energy as mechanical, electrical, thermal or other? What consequence does this have for identifying "heat" as stored energy?

SOLUTION: There is only one "type" of energy, and this quantity is equivalent to mass. Giving energy by-names only marks the problem that with energy alone we cannot distinguish between different processes.

Take a charged capacitor made of two parallel plates. The energy of the capacitor can be increased either by charging it further, or by increasing the separation of the plates. In the former process, energy is added in an electric process while in the latter it is transferred as the result of a mechanical process. There is no way, however, to say how energy was transferred from simply considering the amount of energy stored in the field of the capacitor. Therefore, calling the stored energy "electricity" or "work" would be senseless.

Since "heat" is used as a word for energy transferred in heating, and since the energy of a body such as air can be increased in different ways, calling the energy of a warm body "heat" would be wrong. ■

What happens to all bodies under all circumstances if their energy is increased? Which physical quantity changes if this happens? What kind of conclusions *cannot* be drawn from the statement that the energy of a body has changed?

SOLUTION: The bodies get heavier and more inert, which means that their mass increases.
We cannot conclude, though, that a stone must become hotter or faster from knowing only that its energy has increased.

■

Compare different substancelike physical quantities such as momentum, charge, amount of substance, and entropy. Which two properties do they all have in common? What are possible differences between the quantities listed?

SOLUTION: Their common properties are the following: they all can be stored in systems, and they all can be transferred from system to system (they can flow). Expressed more formally, they all satisfy an equation of balance.

There are some differences, however. For example, momentum and charge are conserved quantities, while amount of substance and entropy are not. There is a further difference between amount of substance and entropy. The former quantity may be produced and destroyed, while the latter may be only produced.
Moreover, momentum and chase may take positive and negative values, while amount of substance and entropy only take positive values. In three dimensions, momentum is a vector, while the other three quantities are scalars.

Naturally, there are other differences as well, giving rise to the variety of phenomena we observe. Charge, for example, gives rise to a field while entropy does not.

■

Consider a box containing water and an immersion heater (Figure 41). Except for the electrical wires leading to the heater, the system is totally insulated from the environment. Let the heater make the water hotter. If you take the entire setup as the system, why would it be wrong to say that the system has been heated? (Remember the definition of heating, or cooling, in Section 1.1.3.) What does this mean with respect to the question of where the heat has come from which has made the water warmer? What is heat in this case?

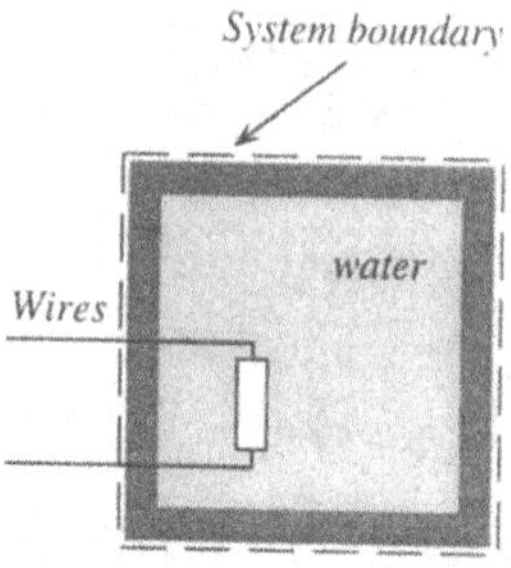

FIGURE 41. Problem 9.

SOLUTION: Heating means that heat (entropy) is flowing into the system. Here, the system is adiabatic, i.e., it is insulated from the surroundings to the flow of heat; therefore, it has not been heated.

If water is to get warmer it necessarily has to increase its heat content. [We neglect the possibility of compression of the almost incompressible substance. See Figure 44 of Chapter 4 for the kind of pressure necessary to appreciably rais the temperature of water by adiabatic compression.] Since heat did not flow in from the outside, the increase of heat must have come from production in the system.

In this description it is impossible for heat to be energy. Remember that energy was added to the system as a consequence of the electric process.

■

An electrical resistor element R_x is used as a thermometer. It is placed as one of the four resistors in a circuit (Figure 42). There is no current through the galvanometer (G) if the thermometer is put in water with a temperature of 20°C and the resistor R_s is given a value of 10.0 kΩ. Now we place the thermometer into a different fluid. Again, electricity does not flow through the galvanometer if we increase R_s by 1000 Ω. What is the temperature of the second fluid? The coefficient of temperature of the platinum resistor is $2.0 \cdot 10^{-3}$ K^{-1}.

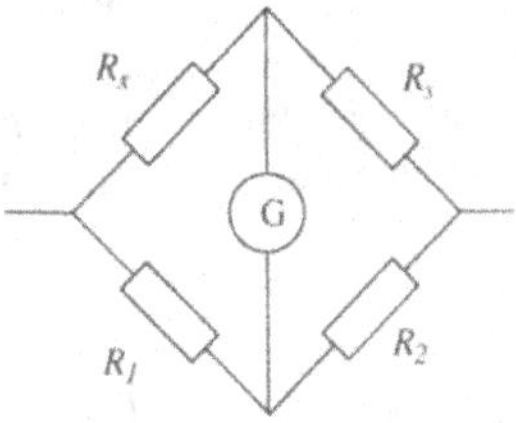

FIGURE 42. Problem 11.

SOLUTION:

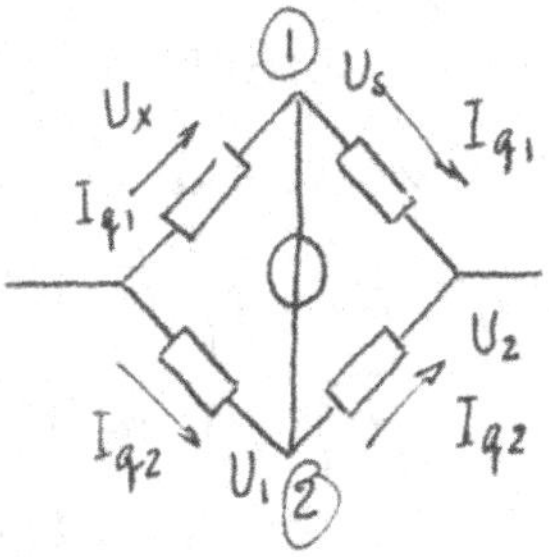

If there is no electric current through the galvanometer, the electric potential at points 1 and 2 must be equal, leading to the conditions

$$U_x = U_1 \quad \text{and} \quad U_s = U_2$$

$$\text{or} \quad R_x I_1 = R_1 I_2 \quad \text{and} \quad R_s I_1 = R_2 I_2$$

Therefore: $$\frac{R_x}{R_s} = \frac{R_1}{R_2} = \text{const}$$

Now: $$\frac{R_{x,new}}{R_{x,20}} = \frac{R_{s,new}}{R_{s,20}} = \frac{11000\ \Omega}{10000\ \Omega} = 1.10$$

Temperature dependence of R_x:

$$R_{x,new} = R_{x,20}(1 + \alpha \Delta T)$$

$$\Delta T = \frac{1}{\alpha}\left(\frac{R_{x,new}}{R_{x,20}} - 1\right) = \frac{1}{2 \cdot 10^{-3}}(1.10 - 1)\ \text{K} = 50\ \text{K} \rightarrow T_{new} = 70°\text{C}$$

■

Do any of the simple processes discussed in Section 1.3 lead from the lower right to the upper left corner of the *T-S* diagram? Could any two achieve the goal of changing a fluid's state in such a manner? Will your answer be different for air and for water in a range of temperatures including the latter's anomaly?

SOLUTION: We are looking at a process which has the following type of curve in the TS-diagram:

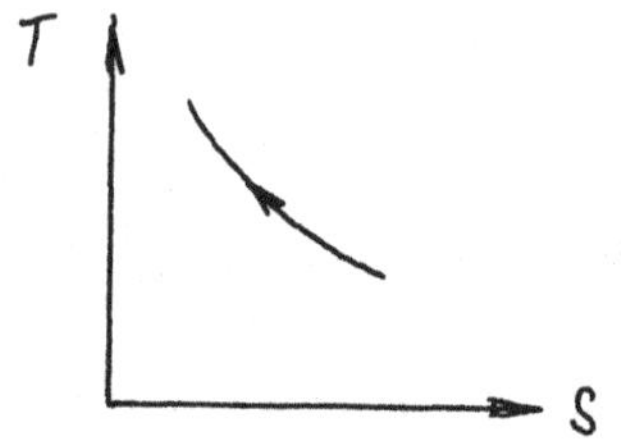

In other words, while the entropy of the system decreases, it's temperature is supposed to increase. Inspection of the TS-diagrams in Section 1.3 (Figures 12-15) shows that the processes do not exhibit such behavior.

Two different processes, such as an isothermal compression (in the case of air) followed by an adiabatic compression can accomplish the stated goal.

In the case of water with temperatures between 0°C and 4°C, we again need two steps such as isothermal and adiabatic changes. This time, however, both processes must <u>increase</u> the volume of the fluid. (See the Interlude, Section I.2.2. for more details.)

■

How large must the entropy current through an ideal Carnot engine be if its power is 10 kW, and the upper and the lower operating temperatures are 500°C and 100°C, respectively? How does your answer change if you double the power? If you double the temperature difference?

SOLUTION :

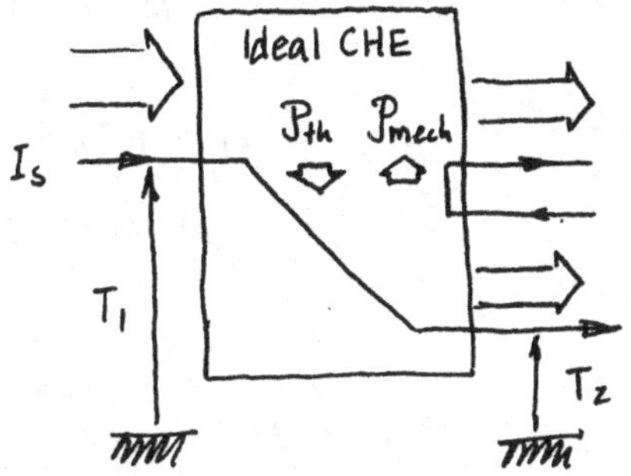

$$|P_{th}| = (T_1 - T_2)\,|I_s|$$

$$|I_s| = \frac{|P_{th}|}{T_1 - T_2}$$

$$= \frac{10000}{773 - 373}\,\frac{W}{K} = 25\,\frac{W}{K}$$

Doubling the power: $|I_s| = 50$ W/K

Doubling the temperature difference:

$$|I_s| = 12.5 \text{ W/K}$$

The analogy using a waterfall gives the same answers.

■

A large power plant of 1.0 GW electrical power emits a thermal energy current to the environment twice as large as the useful one. At what rate does the environment, which has a temperature of 25°C, receive entropy?

SOLUTION:

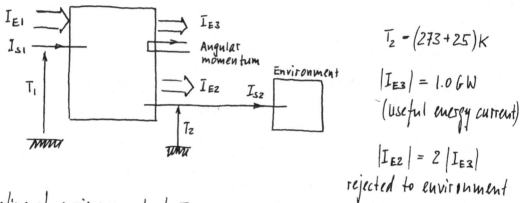

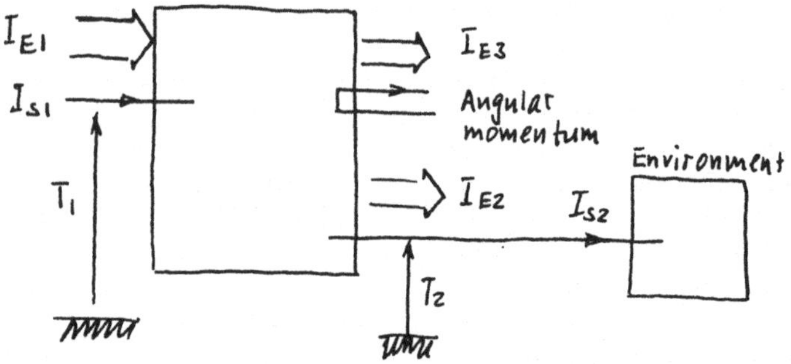

$T_2 = (273+25)\,K$

$|I_{E3}| = 1.0\,GW$ (useful energy current)

$|I_{E2}| = 2\,|I_{E3}|$ rejected to environment

Heating of environment at T_2:

$$|I_{E2}| = T_2\,|I_{S2}|$$

$$\rightarrow |I_{S2}| = \frac{1}{T_2}|I_{E2}| = \frac{1}{298\,K}\,2 \cdot 1.0 \cdot 10^9\,W = 6.7 \cdot 10^6\,\frac{W}{K}$$

I_{S2} is the entropy current entering the environment.

Note: Details of the operation of the heat engine do not have to be known for us to answer this question. For example, it is not necessary that the engine rejects entropy at 25°C. (The temperature will be higher with a heat exchanger placed between the engine and the environment.) The heating of the environment only depends upon its temperature and the energy current received.

■

An immersion heater has a temperature of 120°C as it emits an energy current equal to 0.80 kW. a) How large is the current of entropy flowing across the surface of the heater? b) If the temperature of the water receiving the heat is equal to 80°C, how much entropy flows into the water?

SOLUTION:

a)

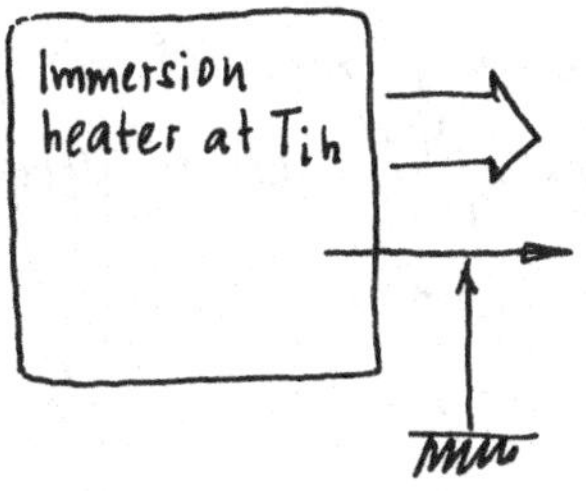

$$|I_s| = |I_E| / T_{ih} = \frac{800\ W}{(120 + 273)\ K} = 2.04\ \frac{W}{K}$$

b)

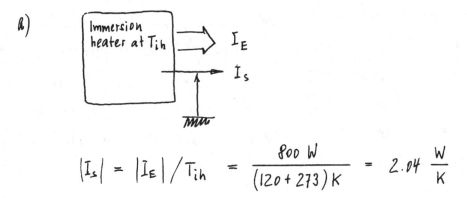

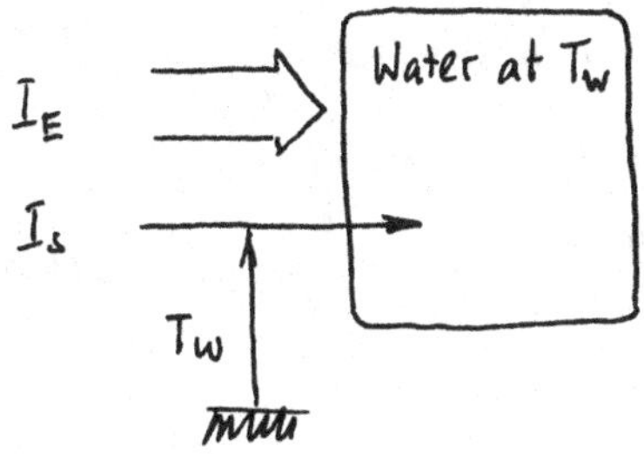

$$|I_s| = |I_E| / T_w = \frac{800\ W}{(80 + 273)\ K} = 2.27\ \frac{W}{K}$$

Obviously, the entropy current increases as it passes from the immersion heater to the water.

■

Sunlight falling on a window is partly reflected (10%) and mostly transmitted (87%). The rest is absorbed evenly throughout the glass. The energy current associated with the radiation is equal to 900 W. If the temperature of the glass is equal to 30°C, how large is the rate at which entropy is received by this body?

SOLUTION: Here, we have to be very careful with words and interpretations, since the absorption of radiation by the glass is irreversible. Indeed, almost all the entropy necessary for increasing the temperature of the glass is produced. (Sunlight carries very little entropy compared to the amount produced by absorption by cool bodies such as those found at the surface of the Earth; see Chapter 3 for more details.) Therefore, we should replace the word "received" in the problem statement by "produced".

However, since it does not matter to the glass whether the entropy appearing in the body was "received" or "produced", we can use the analysis of the last problem (Problem 17). Simply model the production of entropy to take place in a different system not identical with the body of glass:

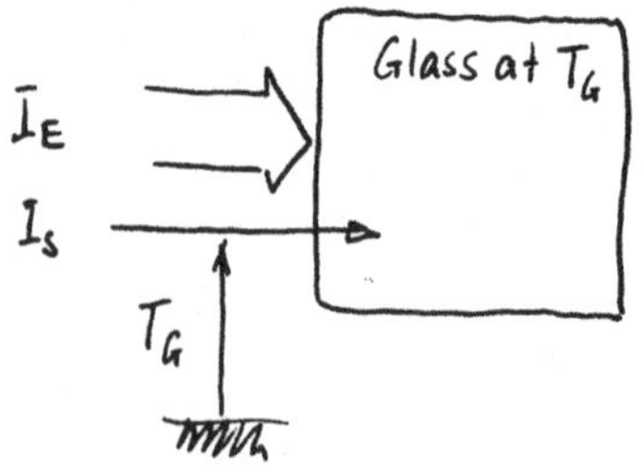

$$|I_S| = |I_E| / T_G = \frac{0.03\,|I_{E,rad}|}{T_G}$$

$$= \frac{0.03 \cdot 900\,W}{(30+273)K} = 0.089\,\frac{W}{K}$$

■

As a body is heated, its current of entropy increases linearly from 20 W/K to 40 W/K in 100 s, and its temperature goes from 100°C to 70°C (also linearly). a) Calculate the thermal energy current received by the body as a function of time. b) How much entropy and energy have been absorbed by the body during the process? c) Is this scenario physically possible? Can a body get colder as it receives heat without losing any?

SOLUTION: c) We should answer this question first. This scenario is indeed possible in a body such as air, if it is expanded while being heated at the same time.

a)

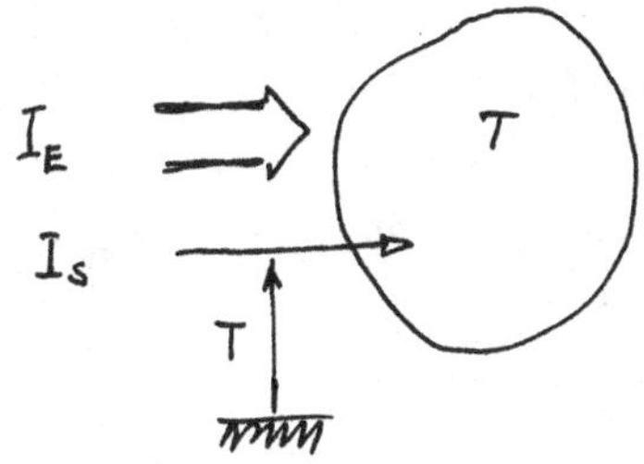

$$T(t) = T_o + at$$

$$T_o = 373\ \text{K}, \quad a = -0.30\ \text{K/s}$$

$$I_s(t) = I_{so} + bt$$

$$I_{so} = -20\ \text{W/K}, \quad b = -0.20\ \text{W/(K·s)}$$

b)

$$S_e = -\int_0^{100s} I_s(t)\,dt = \int_0^{100s}\left[20\ \text{W/K} + 0.20\ \text{W/(K·s)}\cdot t\right]dt$$

$$= 20\ \text{W/K}\cdot 100\text{s} + \frac{1}{2}\,0.20\ \text{W/(K·s)}\,(100\,\text{s})^2 = 3000\ \text{J/K}$$

$$Q = -\int_0^{100s} I_E(t)\,dt = -\int_0^{100s} T(t)\,I_s(t)\,dt$$

$$= -\int_0^{100s}\left[T_o I_{so} + (a I_{so} + b T_o)t + ab\,t^2\right]dt$$

$$= -\left[T_o I_{so} t + \frac{1}{2}(a I_{so} + b T_o)t^2 + \frac{1}{3}ab\,t^3\right]_0^{100s} = 1.07\cdot 10^6\ \text{J}$$

Note: Simply multiplying the average temperature by the average entropy current will yield almost the same numerical answer. ■

A mixture of water and ice is heated in such a way that ice melts at a rate of 0.020 kg/min. a) How large is the current of entropy absorbed by the mixture? b) How much energy does the mixture receive after 10 minutes of heating?

SOLUTION: We model the process of melting of ice as that of a spatially uniform body receiving heat at constant temperature.

a) We need to know the latent entropy of fusion of water. According to Table A.4, this quantity is 22.0 J/(K·mole). Since we are going to work with the mass of the ice we will use the specific value:

$$\lambda_f = \lambda_{fn} / M_o = 22.0/0.018 \ J/(K \cdot kg) = 1.22 \cdot 10^3 \ J/(K \cdot kg)$$

Now we can write Equation (8) in the form

$$I_S = \lambda_f \dot{m}_{ice} = 1.22 \cdot 10^3 \frac{J}{K \cdot kg} \frac{-0.020 \ kg}{60 \ s}$$

$$= -0.41 \ W/K \quad \text{(entropy enters the system)}$$

b) In heating of a system at temperature T we have $I_E = T I_S$ (Equation 13). The energy added in heating is defined by Equation (14):

$$Q = -\int I_E \, dt = -\int T I_S \, dt = -T \int I_S \, dt = -T I_S \Delta t$$

for T = const and I_S = const. With T = 273 K we obtain Q = 66.7 kJ ■

Two or more ideal Carnot heat engines operate in sequence, which means that the entropy rejected by one engine is used by the following one. Each of the engines runs in a distinct interval of temperatures between T_{max} and T_{min}, with the intervals seamlessly covering the entire range of temperatures. a) Show that the power of the sequence of engines is equal to the power of a single engine running between T_{max} and T_{min}. b) Allow for entropy to be added or withdrawn at each inlet to an engine. Show that in this case the power of the sequence of devices should be equal to

$$\mathcal{P} = \sum_{i=1}^{N} I_s(T_i)\Delta T_i$$

$$\Delta T_i = T_i - T_{i+1} \quad , \quad T_1 = T_{\max} \quad , \quad T_{N+1} = T_{\min}$$

c) If the entropy current is a continuous function of temperature (between the maximum and the minimum values), show that the power should be calculated according to

$$\mathcal{P} = \int_{T_{\max}}^{T_{\min}} I_s(T)dT$$

SOLUTION:

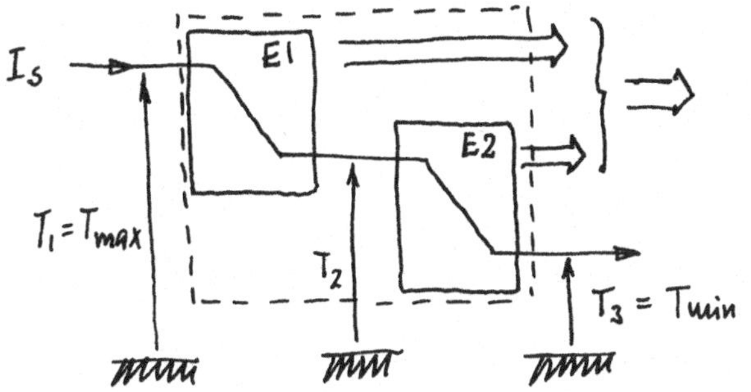

a) We can understand the problem by considering just two engines in sequence:

$$\mathcal{P}_1 + \mathcal{P}_2$$
$$= (T_1 - T_2)|I_s| + (T_2 - T_3)|I_s|$$
$$= (T_1 - T_3)|I_s|$$
$$= (T_{max} - T_{min})|I_s| = \mathcal{P}$$

b) Again consider two engines with an additional current of entropy entering the second one, so that

$$|I_{s2}| = |I_{s1}| + |I_s^*|$$

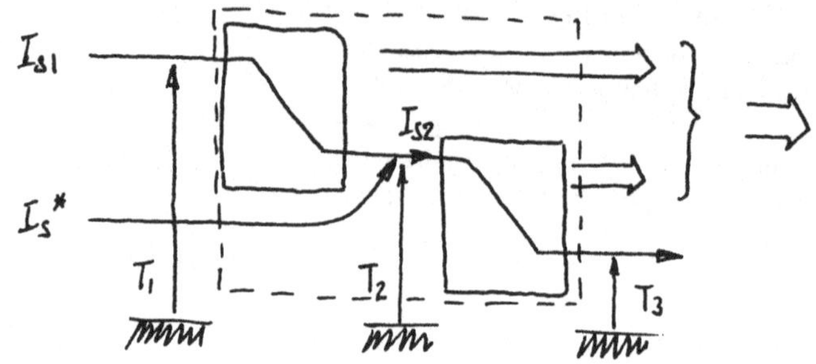

$$\mathcal{P} = \mathcal{P}_1 + \mathcal{P}_2$$
$$= (T_1 - T_2)|I_{S1}| + (T_2 - T_3)|I_{S2}|$$

c) The expression is simply the limiting case for many heat engines with small temperature differences, with a continuous supply of entropy in the entire range of temperatures.

■

Would you treat solar radiation as a high or a low temperature heat source? Discuss the implications of your decision.

SOLUTION:

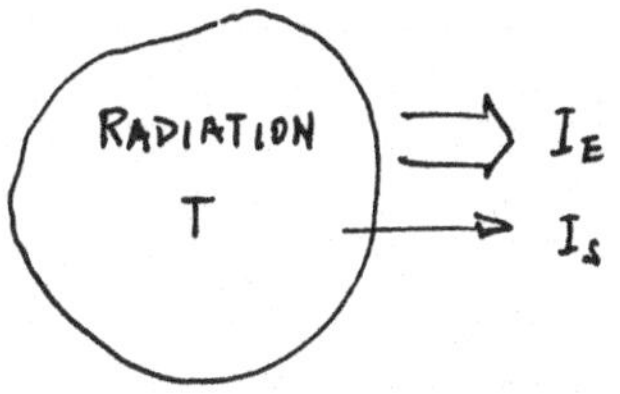

Solar radiation is a heat source, i.e., it carries entropy. It is a high temperature entropy source (near 6000 K), which means that it delivers large quantities of energy per unit amount of entropy; vice-versa, we say that it carries very little entropy per unit amount of energy!

Having a high temperature entropy source combined with a low temperature sink (the Earth) allows for the construction of heat engines with high thermal efficiency. Also, bodies can be heated to high temperatures in concentrated light (to nearly 6000 K).

Had we interpreted solar radiation as a low temperature heat source, these processes could not be explained.

■

The *COP* of a refrigerator is defined as the ratio of the thermal energy current extracted from the cold body and the power needed to drive the engine. a) Derive the formula for the *COP*, and calculate the value for an ideal refrigerator operating between temperatures of – 20°C and 25°C. b) Explain the difference in the viewpoints taken for heat pumps and refrigerators.

SOLUTION:

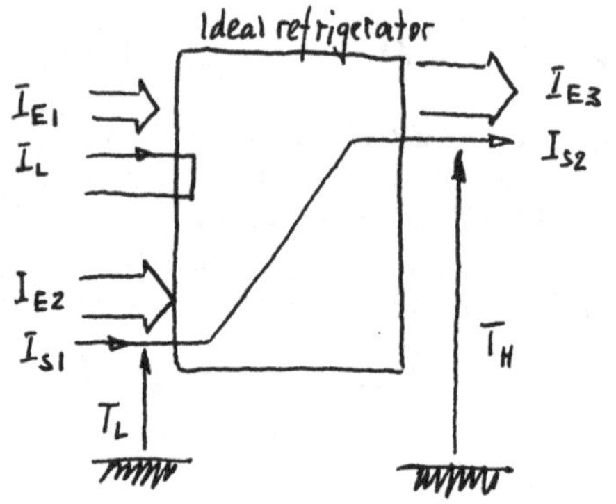

a) Definition of COP:

$$COP = \frac{|I_{E2}|}{|I_{E1}|}$$

Balance of entropy:

$$\Pi_S = 0 \ , \quad |I_{S1}| = |I_{S2}|$$

Energy currents: $|I_{E2}| = T_L |I_{S1}|$, $|I_{E3}| = T_H |I_{S2}|$

$$\Rightarrow \quad COP = \frac{|I_{E2}|}{|I_{E1}|} = \frac{|I_{E2}|}{|I_{E3}| - |I_{E2}|}$$

$$= \frac{T_L |I_{S1}|}{T_H |I_{S2}| - T_L |I_{S1}|} = \frac{T_L}{T_H - T_L}$$

$$COP = \frac{273 - 20}{(273 + 25) - (273 - 20)} = 5.6$$

b) In the case of a heat pump, the energy current leaving the heat pump at high temperature (T_H) is considered instead of the one entering at lower temperature T_L which leads to the formula in Equation (20).

■

A mixture of ice and water is placed in a freezer having a constant interior temperature of – 18°C. If the refrigerator works as an ideal Carnot engine, what is the power needed for its operation? Ice is to be formed at a rate of 10 g per minute. The temperature of the environment is taken to be 22°C. What is the difference between this problem and Example 17?

SOLUTION:

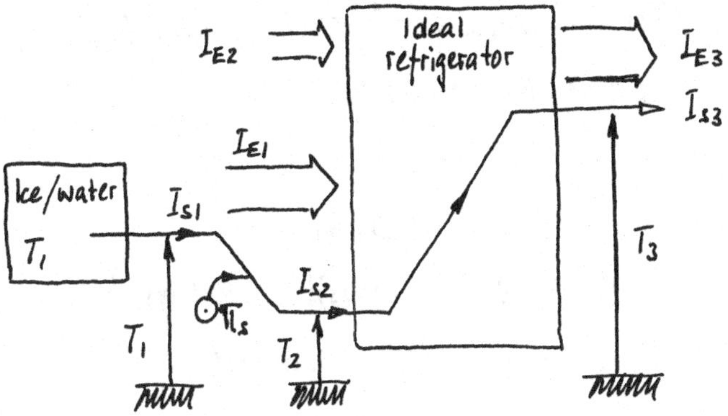

$T_1 = 273\text{ K}$, $T_2 = 255\text{K}$
$T_3 = 295\text{ K}$

Model: The ideal refrigerator operates between T_2 and T_3. Entropy drops from ice/water to T_2 (irreversible).

Energy current leaving ice/water: $|I_{E1}| = q_n/M_o \cdot \dot{m}$
Energy and entropy entering refr.: $|I_{E1}| = T_2 |I_{S2}|$

$$|I_{S2}| = q_n \dot{m} / (T_2 M_o) = \frac{1}{255}\frac{6010}{0.018}\frac{0.010}{60}\frac{\text{W}}{\text{K}}$$

Balance of entropy for refrigerator: $|I_{S2}| = |I_{S3}|$
Power necessary for refrigerator: $|I_{E2}| = (T_3 - T_2)|I_{S2}| = 40\text{K} \cdot 0.218\frac{\text{W}}{\text{K}} = 8.73\text{ W}$

In Example 17 the ideal refrigerator takes up entropy at 273 K, not at 255 K. Moreover, for a given amount of ice formed, there is less entropy to be pumped (no production of entropy). Together these effects lead to twice the energy required for freezing water, as compared to Example 17. ■

In a single stroke of a bicycle pump, air is compressed very rapidly. Compare reversible and dissipative compression. In a dissipative process, would the volume be larger, equal, or smaller than the one found in reversible operation for equal final temperatures? At equal final volume, would the temperature be larger, equal, or smaller in the dissipative case than in the reversible one?

SOLUTION: Rapid compression does not allow for heat (entropy) to flow into or out of the body of air (adiabatic conditions).

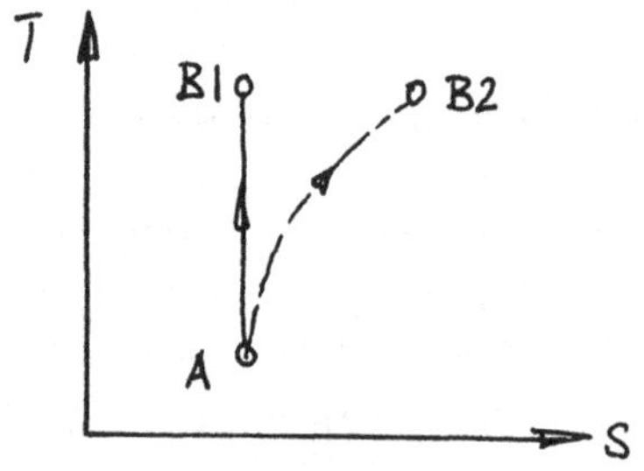

A → B1 : reversible

A → B2 : irreversible, entropy has been produced

In a dissipative process, the final volume would have to be larger at equal final temperature, since more entropy resides in the fluid at the same temperature.

At equal final volume, the entropy content would be larger for the dissipative process, meaning that the temperature would have to be larger too.

■

Over a time span of 100 s, the entropy of a body increases linearly from 300 J / K to 500 J / K. At the same time the rate of generation of entropy decreases from 5 W / K to zero. a) Compute the net flux of entropy as a function of time. b) How much entropy is exchanged, and absorbed, and emitted?

SOLUTION:

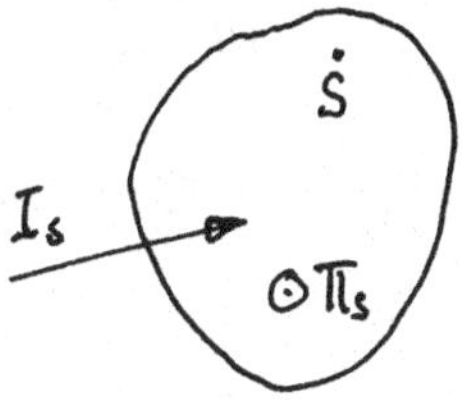

Balance of entropy:

$$\dot{S} = -I_s + \Pi_s$$

$$\dot{S} = \frac{\Delta S}{\Delta t} = \frac{500-300}{100}\frac{W}{K} = 2.0\frac{W}{K}$$

$$\Pi_s = \Pi_{s0} + bt \;, \quad \Pi_{s0} = 5.0\ W/K$$

$$b = -0.050\ W/(K\cdot s)$$

a) $$I_s = \Pi_s - \dot{S}$$

$$= 3.0\frac{W}{K} - 0.050\frac{W}{K\cdot s}\cdot t$$

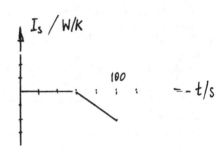

b) $$S_e = -\int_0^{100\,s} I_s(t)\,dt = -50\ J/K$$

$$S_e^+ = \left| \int_{60\,s}^{100\,s} I_s(t)\,dt \right| = 40\ J/K$$

$$S_e^- = \left| \int_0^{60\,s} I_s(t)\,dt \right| = 90\ J/K$$

$$S_e = S_e^+ - S_e^-$$

■

A mixture of ice and water is heated as in Problem 21. a) Calculate the rate of change of entropy of the mixture. b) How large is the rate of change of energy of the volume? Why is it not precisely equal to the energy current in heating?

SOLUTION:

$$\dot{m}_{water} = 3.33 \cdot 10^{-4} \ kg/s$$

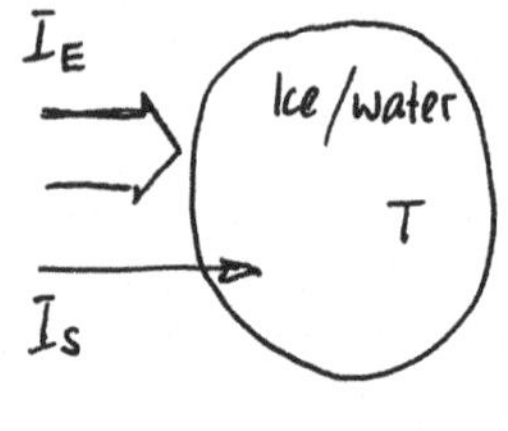

$$a) \ \dot{S} = l_f \dot{m} = \bar{l}_f \dot{n} = \bar{l}_f \frac{1}{M_0} \dot{m}$$

$$= 22.0 \frac{J}{K \cdot mole} \frac{1}{0.018} \frac{mole}{kg} 3.33 \cdot 10^{-4} \frac{kg}{s}$$

$$= 0.41 \ W/K$$

b) Actually, we cannot calculate the quantity asked for with the information available to us. We could calculate the energy current as a result of heating (see below), but this quantity is not equal to the rate of change of the energy of the system. Because of the change of volume (at constant pressure), we also have an exchange of energy in the mechanical process. However, this contribution is small.

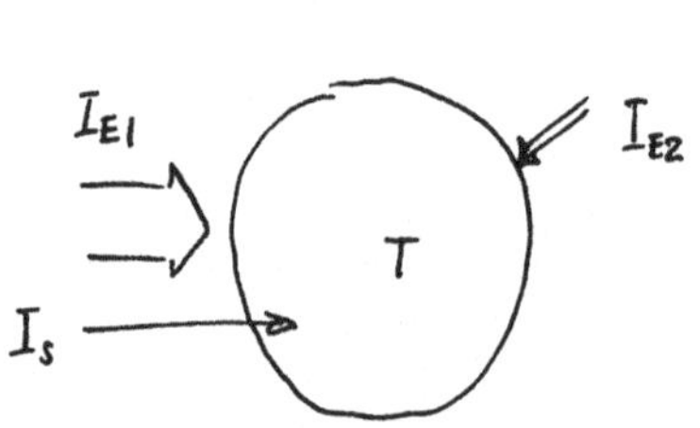

$$I_{E1} = T I_s = T(-\dot{S})$$

$$= 273 \ K \ (-0.41 \ W/K)$$

$$= -111 \ W$$

With $|I_{E2}| \ll |I_{E1}|$, we have $\dot{E} \approx -I_{E1} = 111 \ W$

■

Consider water being heated by an immersion heater. a) If you consider the body of water as a system, what is its equation of balance of entropy? (Assume the distribution of entropy through the system to take place reversibly; what does this mean for the conduction of entropy through the system?) b) Answer the question for the case in which you take the system to be made up of water plus heating coil.

SOLUTION:

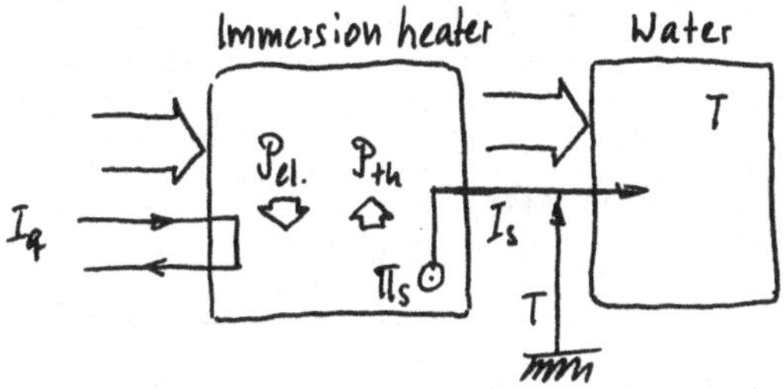

a) The system is made up of the water only.

MODEL: The system is uniform (has the same temperature throughout) and receives entropy from the outside.

Since the entropy does not have to flow from hotter to colder places in the system, it spreads reversibly: $\Pi_{S,water} = 0$. Therefore, the balance of entropy for the system is:

$$\dot{S} = -I_s$$

b)

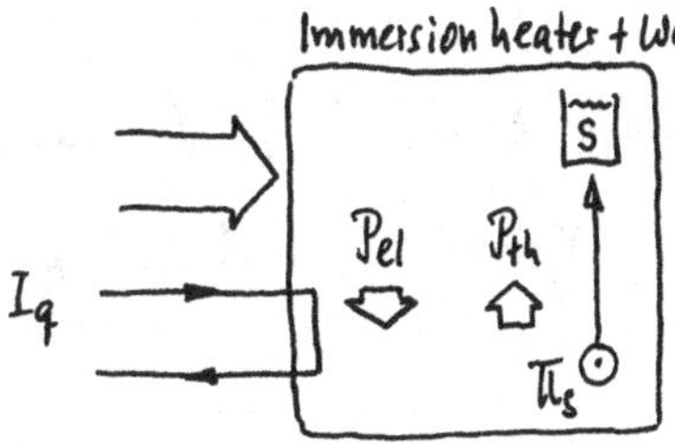

MODEL: The system consisting of water and immersion heater is connected to the environment through the electrical wires only. Therefore, there is no exchange of entropy with the environment. Entropy production takes place inside the system. Balance of entropy:

$$\dot{S} = \Pi_s$$

■

If the isothermal expansion or compression of air is dissipative, is the energy exchanged in heating still equal to the area under the isotherm in the *T-S* diagram? Is there a difference in your answers for compression and expansion?

SOLUTION: In order to calculate the energy exchanged in heating, we need the entropy exchanged:

$$Q = -\int I_{E,th}\, dt = -\int T I_s\, dt = T \cdot S_e$$

The area between the TS-curve and the S-axis, on the other hand, is given by

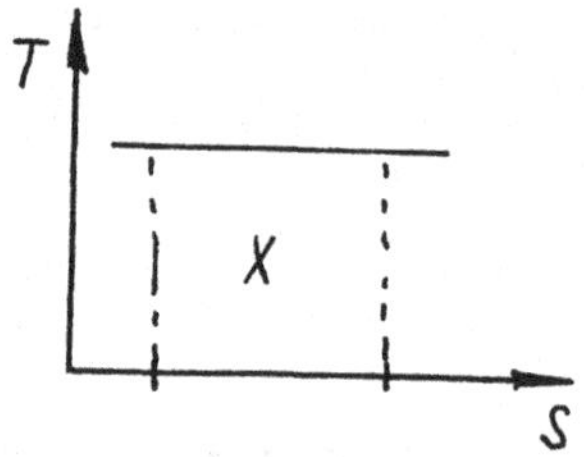

$$X = T\,\Delta S$$

According to the law of balance of entropy, ΔS is equal to

$$\Delta S = S_e + S_{prod}\,.$$

Therefore, $X = T(S_e + S_{prod})$ which is <u>not</u> equal to Q for dissipative processes. The answer is the same for expansion and compression. ∎

Derive the rate of production of entropy for heat transfer which obeys a constitutive law of the form $I_E = a\,(T_1 - T_2)$. a) Write the result in terms of the current of energy. b) Write the formula in terms of the difference of temperatures. c) Two practical problems having to do with heat transfer are thermal insulation and augmentation of transfer. In the former case one wishes to reduce the transfer rate for a given difference of temperatures. In augmentation the rate of transfer is usually prescribed by the problem, and we want to reduce the temperature difference across the heat exchanger. Show that the seemingly contradictory applications are both an exercise in the minimization of the rate of production of entropy.

SOLUTION:

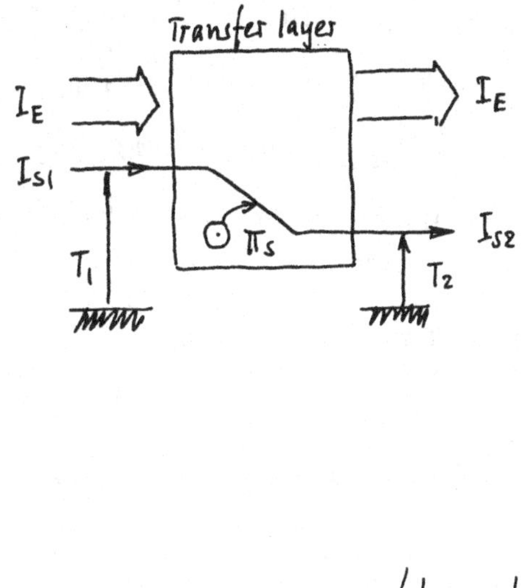

a)
$$I_E = T_1 I_{S1}$$
$$I_E = T_2 I_{S2}$$
$$\to \Pi_S = |I_{S2}| - |I_{S1}|$$
$$= \left(\frac{1}{T_2} - \frac{1}{T_1}\right)|I_E|$$

b)
$$\Pi_S = \frac{T_1 - T_2}{T_1 T_2}|I_E|$$
$$= a\frac{(T_1 - T_2)^2}{T_1 T_2}$$

c) Insulation: $\Pi_S = \left(\frac{1}{T_2} - \frac{1}{T_1}\right)|I_E|$ is reduced if $|I_E|$ is reduced at fixed temperatures T_1 and T_2.

Augmentation: $\Pi_S = \frac{T_1 - T_2}{T_1 T_2}|I_E|$ is reduced if T_2 is increased for fixed $|I_E|$ and T_1.

■

A heat pump is used to heat water at 60°C. Heat is taken from the ground at 2°C. The observed coefficient of performance is 2.2 while the heating power has a magnitude of 1.0 kW. a) How large is the rate of production of entropy? b) How large is the loss of available power? Show that it is equal to the product of the rate of generation of entropy and the temperature of the environment. c) How large is the second law efficiency of the heat pump?

SOLUTION:

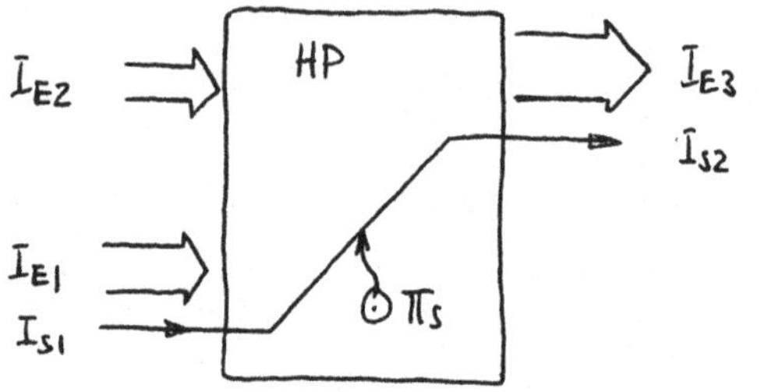

MODEL: All irreversibilities are included with the heat pump.

$|I_{E2}|$ represents the energy current supplied to drive the heat pump. With $|I_{E3}| = 1000$ W and the definition of the COP $= |I_{E3}|/|I_{E2}|$ we can calculate I_{E2}:

$$|I_{E2}| = 1000\,W/2.2 = 455\,W.$$

The balance of energy expressed for the heat pump operating at steady-state delivers the missing energy current:

$$|I_{E1}| = |I_{E3}| - |I_{E2}| = 545\,W$$

a) Balance of entropy: $\Pi_S = |I_{S2}| - |I_{S1}|$

Energy and entropy currents: $|I_{E1}| = T_1\,|I_{S1}|$, $|I_{E3}| = T_2\,|I_{S2}|$

$$\Pi_S = \frac{|I_{E3}|}{T_2} - \frac{|I_{E1}|}{T_1} = \frac{1000W}{333K} - \frac{545W}{275K} = 1.02\,\frac{W}{K}$$

b) The loss of available power for a heat pump is most directly defined as the difference between the actual power and that necessary for driving an ideal heat pump:

$$\mathcal{L} = |I_{E2}| - |I_{E2}|_{ideal}$$

$$COP_{ideal} = T_2/(T_2 - T_1) = 5.74 \Rightarrow |I_{E2}|_{ideal} = 174.2\ W$$

$$\Rightarrow \mathcal{L} = 455\ W - 174\ W = 281\ W$$

Proof of $\mathcal{L} = T_1 \Pi_s$:

$$\mathcal{L} = |I_{E2}| - |I_{E2}|_{ideal} = |I_{E3}| - |I_{E1}| - |I_{E2}|_{ideal}$$

$$|I_{E2}|_{ideal} = (T_2 - T_1)|I_{S2}| \quad \text{(power necessary for pumping } I_{S2}\text{)}$$

$$\Rightarrow \mathcal{L} = T_2|I_{S2}| - T_1|I_{S1}| - (T_2 - T_1)|I_{S2}|$$

$$= T_1\left(|I_{S2}| - |I_{S1}|\right) = T_1 \Pi_s$$

c) The second law efficiency of a thermal engine is the ratio of ideal and actual power (or its inverse):

$$\eta_2 = \frac{|I_{E2}|_{ideal}}{|I_{E2}|} = \frac{174\ W}{455\ W} = 0.38$$

This is also equal to the ratio of the real coefficient of performance and the ideal COP.

■

Derive the formulas for the coefficient of performance and the second law efficiency of a dissipative refrigerator operating between a cold space at temperature T and the environment at T_a in terms of the rate of production of entropy.

SOLUTION:

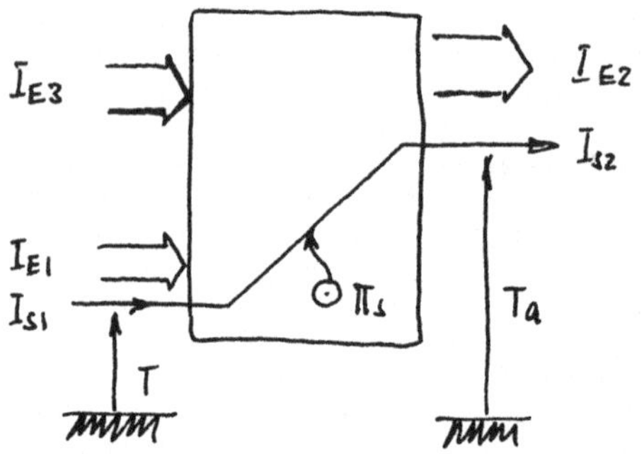

Balance of entropy:

$$|I_{S2}| = |I_{S1}| + \Pi_s$$

Balance of energy:

$$|I_{E2}| = |I_{E1}| + |I_{E3}|$$

Currents: $|I_{E1}| = T|I_{S1}|$; $|I_{E2}| = T_a|I_{S2}|$

Coefficient of performance:

$$COP = \frac{|I_{E1}|}{|I_{E3}|} = \frac{T|I_{S1}|}{T_a|I_{S2}| - T|I_{S1}|} = \frac{T|I_{S1}|}{T_a|I_{S1}| + T_a\Pi_s - T|I_{S1}|}$$

$$= \frac{T}{T_a - T + \frac{T_a}{|I_{S1}|}\Pi_s}$$

Second law efficiency (Equations (49) and (42)):

$$\eta_2 = \left|\frac{\mathcal{P}_{av}}{\mathcal{P}_{mech}}\right| = \frac{(T_a - T)|I_{S1}|}{(T_a - T)|I_{S1}| + T_a\Pi_s} = \frac{T_a - T}{T_a - T + \frac{T_a}{|I_{S1}|}\Pi_s} < 1$$

(The same result can be obtained by calculating COP/COP_{ideal}.) ■

One kilogram of ice is formed from water at 0°C in a freezer. The fluid of the engine operates between – 20°C and 30°C, with the temperature of the environment being held at 20°C. Assume the engine to work reversibly. a) Draw a diagram showing the hotness levels involved and the flow of entropy. Identify sources of entropy production. b) How much entropy is generated? c) How much energy is used for driving the engine in excess of what would be necessary if the engine could operate directly between 0°C and 20°C? d) Verify numerically that the excess work (lost available work) is given by the product of the entropy created and the temperature of the environment. e) Verify formally the Gouy-Stodola rule for lost work.

SOLUTION: a)

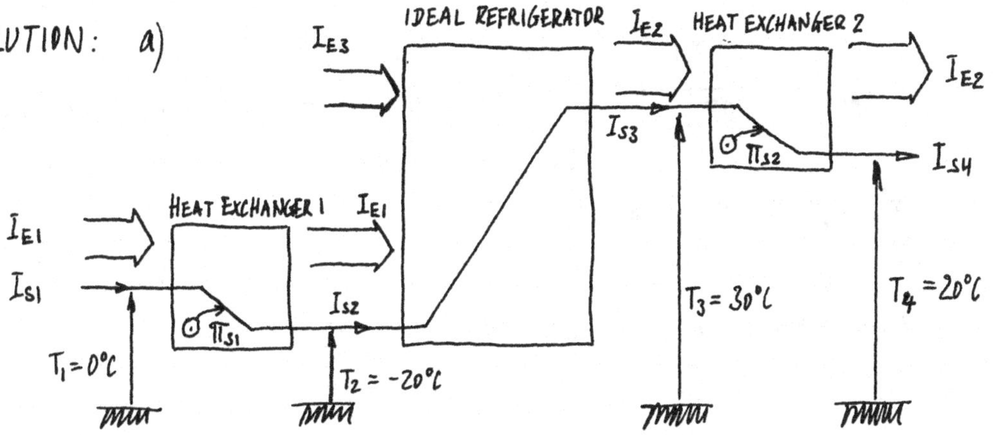

$$Q_1 = \left| \int I_{E1}\, dt \right| \;;\quad Q_2 = \left| \int I_{E2}\, dt \right| \;;\quad W = \left| \int I_{E3}\, dt \right|$$

Entropy is produced in the heat exchangers.

b) Energy transfer: $|Q_1| = q \cdot m = 334 \cdot 10^3\ \text{J/kg} \cdot 1.0\ \text{kg} = 334\ \text{kJ}$

This amount of energy is emitted by the freezing water and must be transferred to the environment.

$$S_{prod,1} = \left(\frac{1}{T_2} - \frac{1}{T_1}\right)|Q| = \left(\frac{1}{253} - \frac{1}{273}\right) 334 \cdot 10^3\ \text{J/K} = 97\ \text{J/K}$$

$$|Q_2| = T_3\,|S_{e3}| = T_3\,|S_{e2}| = T_3\left(|S_{e1}| + S_{prod,1}\right)$$
$$= T_3\left(|Q_1|/T_1 + S_{prod,1}\right) = 303\cdot\left(334\cdot 10^4/273 + 97\right)\,\mathrm{J} = 400\,\mathrm{kJ}$$

$$S_{prod,2} = \left(\frac{1}{T_4} - \frac{1}{T_3}\right)|Q_2| = \left(\frac{1}{293} - \frac{1}{303}\right)400\cdot 10^3\,\mathrm{J/K} = 45\,\mathrm{J/K}$$

c) $$|W_{real}| = |Q_2| - |Q_1| = 66\,\mathrm{kJ}$$

$$|W_{ideal}| = \frac{|Q_1|}{COP_{ideal}} = \frac{T_4 - T_1}{T_1}\,|Q_1| = \frac{20}{273}\,334\cdot 10^3\,\mathrm{J} = 24.5\,\mathrm{kJ}$$

$$|W_{excess}| = 41.5\,\mathrm{kJ}$$

d) $$|W_{loss}| = T_4\,S_{prod} = 293\,(97 + 45)\,\mathrm{J} = 41.6\,\mathrm{kJ}$$

e) $$|W_{loss}| = |W_{real}| - |W_{ideal}|$$
$$= (T_3 - T_2)\,|S_{e2}| - (T_4 - T_1)\,|S_{e1}|$$
$$= T_3\,|S_{e2}| - |Q_1| - T_4|S_{e1}| + |Q_1|$$
$$= T_3\,|S_{e2}| - T_4\,|S_{e1}|$$
$$= |Q_2| - T_4\left(|S_{e4}| - S_{prod}\right)$$
$$= |Q_2| - |Q_2| + T_4\,S_{prod} = T_4\,S_{prod}$$

■

Compare two methods of heating 50°C water. (1) In the first, water at a temperature of 50°C is heated further by a solar collector. (2) The radiation of the sun is used to drive an ideal Carnot heat engine between temperatures of 300°C and 15°C. The energy released by this ideal power plant is used to drive an ideal heat pump which gets its entropy at 15°C to heat the 50°C water. In both cases, 50% of solar radiation is utilized. a) How large is the ratio of the efficiencies of the methods? b) Determine the ratio of the rates of production of entropy for the methods. Assume solar radiation not to supply any entropy.

SOLUTION:

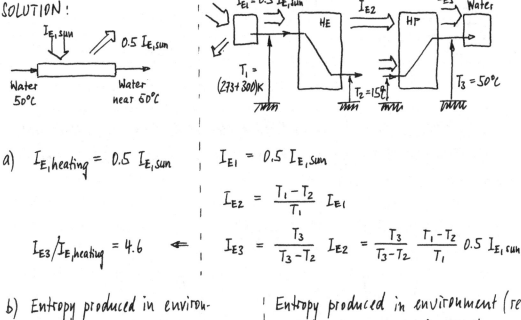

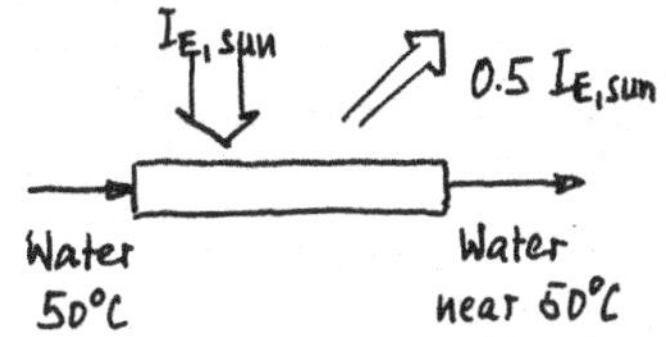

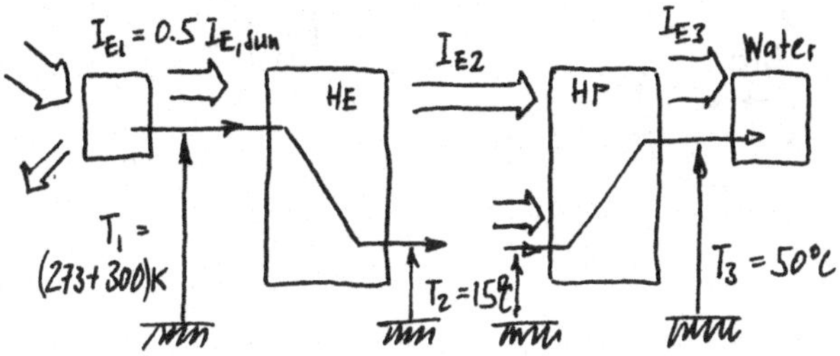

a) $I_{E,heating} = 0.5\, I_{E,sun}$

$$I_{E1} = 0.5\, I_{E,sun}$$

$$I_{E2} = \frac{T_1 - T_2}{T_1} I_{E1}$$

$$I_{E3} = \frac{T_3}{T_3 - T_2} I_{E2} = \frac{T_3}{T_3 - T_2}\,\frac{T_1 - T_2}{T_1}\, 0.5\, I_{E,sun}$$

$I_{E3} / I_{E,heating} = 4.6$ ⇐

b) Entropy produced in environment (rejected solar radiation) and in water:

$$\Pi_{S1} = \Pi_{S,e} + \Pi_{S,w}$$

$$= \frac{1}{T_e} 0.5\, I_{E,sun} + \frac{1}{T_w} 0.5\, I_{E,sun}$$

$$= \left(1.74 \cdot 10^{-3} + 1.55 \cdot 10^{-3}\right) I_{E,sun}$$

Entropy produced in environment (rejected solar radiation). All other operations are reversible:

$$\Pi_{S2} = \Pi_{S,e}$$

$$= \frac{1}{T_e} 0.5\, I_{E,sun}$$

$$= 1.74 \cdot 10^{-3}\, I_{E,sun}$$

$$\Pi_{S1} / \Pi_{S2} = 1.89$$

■

A liter of water is to be frozen in a freezer. Consider only the process of freezing. a) How much entropy is emitted by the water during this process? b) What is the minimal amount of energy which you have to supply to the freezer if the entropy is to be rejected to the environment at 22°C?

SOLUTION:

a)

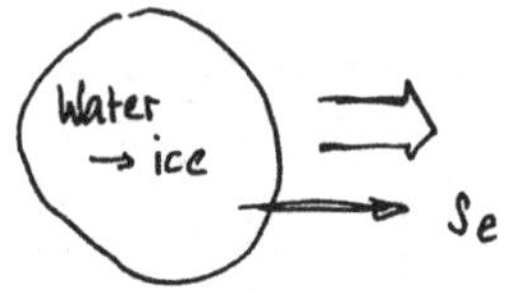

$$|S_e| = l_f\,|\Delta m|$$

$$= 1220\,\frac{\mathrm{J}}{\mathrm{K\cdot kg}}\;1.0\,\mathrm{kg} = 1220\,\frac{\mathrm{J}}{\mathrm{K}}$$

b) Minimal amount of energy: Energy necessary to drive an ideal heat pump (refrigerator) to pump the entropy emitted by the freezing water from 0°C to 22°C:

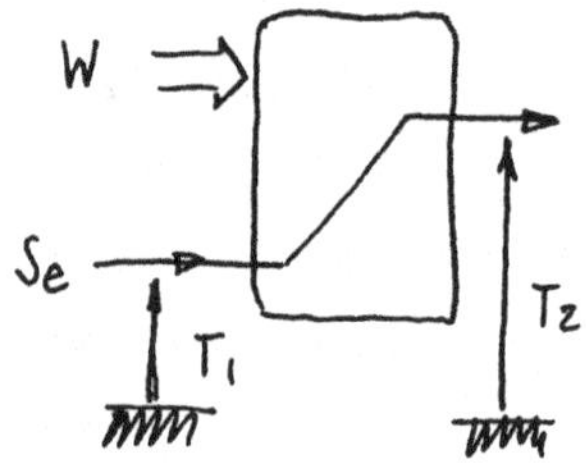

$$|W| = (T_2 - T_1)\,|S_e|$$

$$= 22\,\mathrm{K}\cdot 1220\,\frac{\mathrm{J}}{\mathrm{K}}$$

$$= 26.8\ \mathrm{kJ}$$

■

An endoreversible engine as in Section 1.7 is to be designed. It consists of the reversible Carnot engine and two heat exchangers serving the furnace and the cooler.

a) Prove that its power at maximum output can be written as

$$I_{E,mech} = \frac{(hA)_f}{1+(hA)_f/(hA)_c} T\left[1-\sqrt{\frac{T_o}{T}}\right]^2$$

where f and c refer to the furnace and the cooler, respectively.

b) If the power is maximized once more by dimensioning relative sizes of the heat exchangers optimally, we get

$$I_{E,mech} = \frac{1}{4} hA \cdot T\left[1-\sqrt{\frac{T_o}{T}}\right]^2$$

where $hA = (hA)_f + (hA)_c$ is the total transfer coefficient multiplying the difference of temperatures.

c) Show that the optimized power of such an engine increases proportionally to $(T-T_o)^2/T$ for differences of temperatures which are not too large. This condition is quite applicable to today's range of temperatures. What does this mean for the designer of a power plant?

SOLUTION:

a) $$\left|I_{E,mech}\right|_{max} = \eta_{max} \left|I_{E,in}\right|_{max}$$

Equation (61): $$\left|I_{E,in}\right|_{max} = \frac{(hA)_f \cdot (hA)_c}{(hA)_f + (hA)_c} T\left[1-\sqrt{T_o/T}\right]$$

Equation (63): $$\eta_{max} = 1-\sqrt{T_o/T}$$

$$\Rightarrow \left|I_{E,mech}\right|_{max} = \frac{(hA)_f}{1+(hA)_f/(hA)_c} T\left[1-\sqrt{T_o/T}\right]^2$$

b) $$|I_{E,mech}|_{max} = b \cdot \frac{(hA)_f}{1 + \frac{(hA)_f}{hA - (hA)_f}} \quad ; \quad b = T\left[1-\sqrt{T_0/T}\right]^2$$

$$\Rightarrow \quad |I_{E,mech}|_{max} = b\,\frac{(hA)_f\,(hA - (hA)_f)}{hA}$$

$$\frac{\partial |I_{E,mech}|_{max}}{\partial (hA)_f} = \frac{b}{hA}\left(hA - 2(hA)_f\right)$$

$$\frac{\partial |I_{E,mech}|_{max}}{\partial (hA)_f} = 0 \quad \rightarrow \quad (hA)_{f,\,opt.} = \frac{1}{2} hA$$

$$\Rightarrow \quad |I_{E,mech}|_{opt.} = b\,\frac{1/2\,hA}{1+1} = \frac{1}{4} hA\, T\left[1-\sqrt{T_0/T}\right]^2$$

c) $$|I_{E,mech}|_{opt.} = \frac{1}{4} hAT\left[1-\sqrt{T_0/T}\right]^2$$

$$= \frac{1}{4} hAT\left[1-\sqrt{\frac{T-(T-T_0)}{T}}\right]^2$$

$$= \frac{1}{4} hAT\left[1-\sqrt{1-\frac{\Delta T}{T}}\right]^2$$

$$\approx \frac{1}{4} hAT\left[1-\left(1-\frac{1}{2}\frac{\Delta T}{T}\right)\right]^2 = \frac{1}{4} hAT\left[\frac{1}{2}\frac{\Delta T}{T}\right]^2$$

$$= \frac{1}{16} hA\,\frac{\Delta T^2}{T}$$

■

The furnace of a large thermal power plant was designed to deliver energy at a rate of up to 2.0 GW at a temperature of 920 K. Cooling is done at an environmental temperature of 300 K. Model the engine as endoreversible. a) How large is the current of entropy entering the system? b) What is the optimal mechanical power if heat leakage is responsible for a loss of 5% of the heating power? c) What are the magnitudes of the rate of production of entropy and of the loss of available power? Are they related by the Gouy-Stodola rule?

SOLUTION:

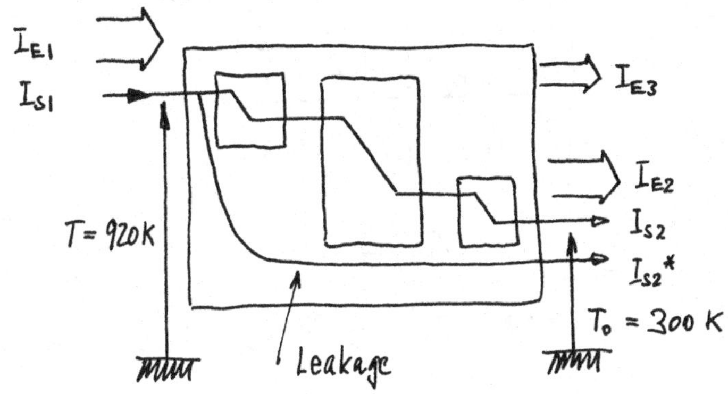

a)
$$|I_{S1}| = |I_{E1}|/T = \frac{2.0 \cdot 10^9}{920}\,\frac{W}{K} = 2.174 \cdot 10^6\ W/K$$

b)
$$|I_{E3}| = \eta_{max}\, 0.95 |I_{E1}| = \left(1 - \sqrt{\frac{300}{920}}\right) 0.95 \cdot 2 \cdot 10^9\ W = 815\ MW$$

c)
$$|I_{S2}| + |I_{S2}^*| = \frac{1}{T_o}\left(|I_{E1}| - |I_{E3}|\right) = \frac{1}{300}\left(2.0 \cdot 10^9 - 8.15 \cdot 10^8\right)\frac{W}{K} = 3.95 \cdot 10^6\ W/K$$

$$\Pi_S = |I_{S2}| + |I_{S2}^*| - |I_{S1}| = 1.776 \cdot 10^6\ W/K$$

$$\mathcal{L} = |I_{E3}|_{ideal} - |I_{E3}| = (T - T_o)|I_{S1}| - |I_{E3}|$$
$$= 620 \cdot 2.17 \cdot 10^6\ W - 815 \cdot 10^6\ W = 5.33 \cdot 10^8\ W$$

Gouy-Stodola: $\mathcal{L} = T_o \Pi_S = 300 \cdot 1.776 \cdot 10^6\ W = 5.33 \cdot 10^8\ W$

■

Derive a general expression for the second law efficiency of an endoreversible engine. Show that it is given by

$$\eta_2 = \frac{1}{1+\sqrt{T_o/T}}$$

SOLUTION: According to definition, $\eta_2 = |I_{E,mech}| / |I_{E,mech}|_{ideal}$

$$\eta_2 = \frac{|I_{E,mech}|}{|I_{E,mech}|_{ideal}} = \frac{|I_{E,heating}|\left(1-\sqrt{T_0/T}\right)}{|I_{E,heating}|\,(T-T_0)/T}$$

$$= \frac{1-\sqrt{T_0/T}}{1-T_0/T} = \frac{1-\sqrt{T_0/T}}{\left(1-\sqrt{T_0/T}\right)\left(1+\sqrt{T_0/T}\right)}$$

$$\Rightarrow \quad \eta_2 = \frac{1}{1+\sqrt{T_0/T}}$$

Note: The second law efficiency of an endoreversible engine lies between 0.5 and 1. For very high upper temperature, the engine comes close to $\eta_2 = 1$.

■

Model a refrigerator as an endoreversible engine. Its purpose is to pump heat at a prescribed rate out of the cold enclosure. The heat exchangers at the colder and at the warmer end have been dimensioned so as to make the temperature differences across them roughly equal. You now can add a piece of heat exchanger to only one of the existing exchangers. Which one do you choose? Assume the temperature differences across the heat exchangers to be small compared to the temperatures themselves and to the difference of the temperatures of the cold enclosure and the environment. The added piece of heat exchanger is small compared to the existing ones.

SOLUTION:

MODEL

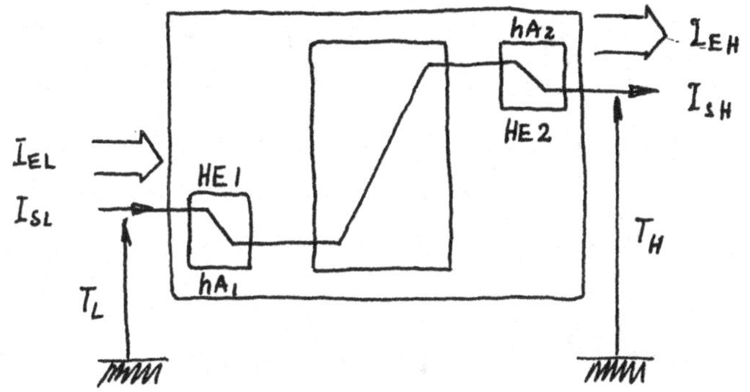

I_{SL}, T_L, and T_H are fixed. Since the temperature differences across the heat exchangers are supposed to be equal and small, we have

$$\frac{|I_{EH}|}{|I_{EL}|} \approx \frac{T_H}{T_L} \quad , \quad \frac{hA_2}{hA_1} \approx \frac{T_H}{T_L}$$

Optimizing thermal processes means minimizing entropy production. Therefore we will investigate which of the alternatives of adding heat exchanger area will lead to a larger decrease of entropy production. The entropy production in a heat exchanger is given by

$$\Pi_s = \left(\frac{1}{T_2} - \frac{1}{T_1}\right)|I_E| = \frac{T_1 - T_2}{T_1 T_2}|I_E| = \frac{1}{hA}\frac{|I_E|^2}{T_1 T_2}$$

$$\approx \frac{1}{hA}\frac{|I_E|^2}{T^2}$$

where T_1 and T_2 are the upper and lower temperatures for a heat exchanger ($T_1 \approx T_2 = T$). Adding a piece of heat exchanger leads to:

$$\Pi_{s,new} = \frac{1}{hA + \Delta hA}\frac{I_E^2}{T^2} \approx \left(1 - \frac{\Delta hA}{hA}\right)\frac{1}{hA}\frac{I_E^2}{T^2}$$

$$\Rightarrow \quad \Delta\Pi_s = -\frac{\Delta hA}{(hA)^2}\frac{I_E^2}{T^2}$$

This can be applied separately to both heat exchangers. The factor $\Delta hA \cdot I_E^2/T^2$ is nearly equal for both, therefore, $|\Delta\Pi_s|$ is larger for the heat exchanger with smaller hA-value, which is the one at the <u>cold</u> end.

■

Consider a solar furnace with heat engine and heat pump as in Example 18. Water is to be heated at constant temperature T_W. The engines are supposed to be ideal Carnot engines. Heat (entropy) is supplied to and rejected from engines and reservoirs at constant temperatures. Assume the furnace to be an ideal absorber of solar radiation. The losses from the furnace to the environment are taken to be proportional to the difference of temperatures, Equation (50), with constant heat transfer coefficient h. a) Derive the expression for the rate of production of entropy for the entire system. b) How large should the temperature of the furnace be to minimize the production rate of entropy? c) For which value of the temperature of the furnace will the power of the heat engine be a maximum? Why is this value different from the one computed in (b)? Should you try to minimize entropy production or to maximize power output of the heat engine? d) Show that the maximum of the total heating power with which the body of water is being heated occurs at the same temperature of the furnace as that calculated for minimal entropy production.

SOLUTION :

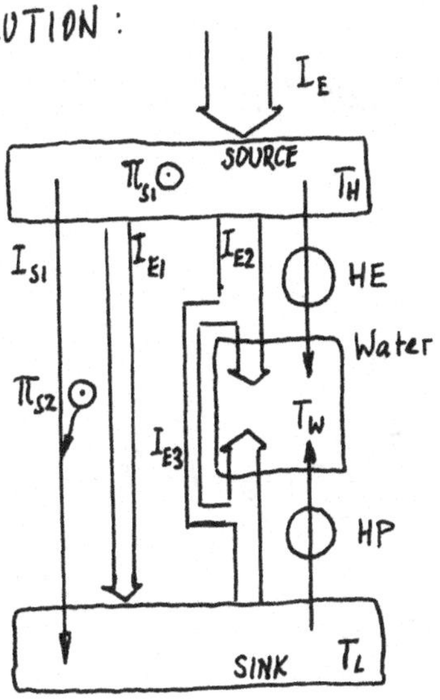

a) Entropy is produced as a consequence of absorption of light in the source (body at T_L) and of transport of entropy from the source to the sink; all other operations are reversible.

$$\Pi_S = \Pi_{S1} + \Pi_{S2} = \frac{I_E}{T_H} + \left(\frac{1}{T_L} - \frac{1}{T_H}\right) I_{E1}$$

$$= \frac{1}{T_H}\left(I_E - I_{E1}\right) + \frac{1}{T_L} I_{E1}$$

$$= \frac{1}{T_H}\left(I_E - hA\left(T_H - T_L\right)\right) + \frac{1}{T_L} hA\left(T_H - T_L\right)$$

$$\rightarrow \quad \Pi_S = \frac{1}{T_H} I_E - hA\left(\frac{T_H - T_L}{T_H} - \frac{T_H - T_L}{T_L}\right)$$

$$= \frac{1}{T_H} I_E - hA\left(2 - \frac{T_L}{T_H} - \frac{T_H}{T_L}\right)$$

b) Rate of production of entropy has to be minimal:

$$\frac{\partial \Pi_S}{\partial T_H} = -\frac{1}{T_H^2} I_E - hA\left(\frac{T_L}{T_H^2} - \frac{1}{T_L}\right)$$

$$\frac{\partial \Pi_S}{\partial T_H} = 0 \Rightarrow \quad -I_E - hA\left(T_L - \frac{T_H^2}{T_L}\right) = 0$$

$$T_L I_E - hA\left(T_H^2 - T_L^2\right) = 0$$

$$T_H = \sqrt{\frac{T_L I_E}{hA} + T_L^2}$$

For $T_L = 293$ K, $h = 2$ W/(K·m²), and $I_E/A = 1000$ W/m², the result yields a numerical value of $T_H = 482$ K

c) Power of the heat engine corresponds to I_{E3}

$$I_{E3} = \frac{T_H - T_W}{T_H} I_{E2} = \frac{T_H - T_W}{T_H}\left(I_E - I_{E1}\right) = \frac{T_H - T_W}{T_H}\left(I_E - hA\left(T_H - T_L\right)\right)$$

$$\frac{\partial I_{E3}}{\partial T_H} = \frac{T_H - T_W}{T_H}(-hA) + \frac{T_W}{T_H^2}\left(I_E - hA\left(T_H - T_L\right)\right)$$

$$\frac{\partial I_{E3}}{\partial T_H} = 0 \Rightarrow \quad -hA\,T_H\left(T_H - T_W\right) + T_W\left(I_E - hA\left(T_H - T_L\right)\right) = 0$$

$$\Rightarrow \quad T_H = \sqrt{\frac{1}{hA} T_W \left(I_E + hA\,T_L\right)} = 514\text{ K} \quad (\text{for } T_W = 333\text{K})$$

Reason for difference: There are two sinks for entropy at T_L and T_W.

d) The water is heated by two entropy currents which together deliver energy at the following rate:

$$\text{heating power} = \left(I_{E2} - I_{E3}\right) + \frac{T_W}{T_W - T_L} I_{E3}$$

$$= \left(I_E - hA\left(T_H - T_L\right) - I_{E3}\right) + \frac{T_W}{T_W - T_L} I_{E3}$$

where $I_{E3} = \frac{T_H - T_W}{T_H}\left(I_E - hA\left(T_H - T_L\right)\right)$ (see Problem c)

$$\text{heating power} = I_E - hA\left(T_H - T_L\right) - \frac{T_H - T_W}{T_H}\left(I_E - hA\left(T_H - T_L\right)\right) + \frac{T_W}{T_W - T_L}\,\frac{T_H - T_W}{T_H}\left(I_E - hA\left(T_H - T_L\right)\right)$$

$$= \left[1 - \frac{T_H - T_W}{T_H} + \frac{T_W}{T_H}\,\frac{T_H - T_W}{T_W - T_L}\right]\left(I_E - hA\left(T_H - T_L\right)\right)$$

$$\frac{\partial P_{heating}}{\partial T_H} = \left[-\frac{T_W}{T_H^2} - \frac{T_W}{T_H^2}\,\frac{T_H - T_W}{T_W - T_L} + \frac{T_W}{T_H}\,\frac{1}{T_W - T_L}\right]\left(I_E - hA\left(T_H - T_L\right)\right) + \left[1 - \frac{T_H - T_W}{T_H} + \frac{T_W}{T_H}\,\frac{T_H - T_W}{T_W - T_L}\right]\left(-hA\right)$$

$$= \frac{T_W T_L}{T_H^{\,2}\left(T_W - T_L\right)}\left(I_E - hA\left(T_H - T_L\right)\right) - \frac{T_W T_H - T_W T_L}{T_H\left(T_W - T_L\right)}hA$$

$$\frac{\partial P_{heating}}{\partial T_H} = 0 \quad \Rightarrow \quad T_L I_E - hA\left(-T_L^2 + T_H^2\right) = 0 \quad \text{(see Problem a).}$$

■

A low temperature heat engine employing a traditional coolant such as R 134 is designed for use with normal solar collectors. Estimate the efficiency you might expect from such an engine if heat is collected at 90°C and rejected at 30°C. (A detailed calculation carried out for a particular design gives a value of 9% to 10% for the thermal efficiency. T. Koch, Diploma thesis 1993, Technikum Winterthur.)

SOLUTION: The best estimate of the thermal efficiency of a real heat engine including the heat exchangers available to us so far is provided by the Curzon-Ahlborn formula (Equation 63) which yields

$$\eta_{CA} = 1 - \sqrt{\frac{T_0}{T}} = 1 - \sqrt{\frac{303}{363}} = 0.09$$

which compares very well with more detailed calculations. However, this neglects the thermal efficiency of the solar collectors which may be 50% or better (with vacuum tube collectors; see Chapter 4.6 for more details on solar collectors). All in all, therefore, the thermal efficiency of such an engine is around 5% measured with respect to the energy of solar radiation. An efficiency around 10% could be obtained if the collectors could be used to directly heat the working fluid.

■

Compare the statements "By itself heat only flows from hotter to colder places" and "By itself water only flows downhill." Why would one statement be considered something truly noteworthy (the second law of thermodynamics) while the other is taken to be almost trivial? In your opinion do the statements suggest a comparison between heat and water, or between heat and energy?

SOLUTION: Consider in turn the two interpretations of heat.

1. Heat is entropy. In this picture the fact that heat by itself only flows "downhill" is a natural consequence of the properties of substancelike quantities (such as water) and their relation to energy. When entropy flows "downhill", energy is released. If it is to be made to flow "uphill", energy is needed.

2. Heat is energy. Since energy is not generally known to have the property of flowing only "downhill" by itself, the observation appears to be noteworthy. Also, from mechanics we do not necessarily learn that energy can flow. Therefore, the statement about heat flowing "downhill" adds new aspects to our knowledge of the physical world (as long as we do not know about entropy). When investigated in detail, however, the observation simply leads to the conclusion that there exists a new quantity whose properties are responsible for energy to flow "downhill" in conduction of heat. (See, for example, D.R. Owen: A First Course in the Mathematical Foundations of Thermodynamics. Springer-Verlag, New York, 1984.) Naturally, this quantity is entropy.

The observation that heat, by itself, flows only "downhill" is one of many different concrete phenomena resulting from the existence and the properties of entropy. Using one of them as the statement of a fundamental law (the second law of thermodynamics) seems tenuous at best. Indeed, traditionally, many different special forms of the second law have been used (see Problem 61) leading to the question which of them could "truly" represent the fundamental law.

In summary, in my personal view, observations such as "by itself heat only flows downhill" should directly be taken as evidence for the existence of a substance like thermal quantity akin to water in hydraulic phenomena.

■

CHAPTER 2

Solutions of Selected Problems

Explain the meaning of uniform processes. How must heat spread through a body for such a process to be reversible? Is a uniform process necessarily reversible?

SOLUTION: A uniform process is one during which quantities such as temperature, density, and pressure have the same value throughout the body at a given moment; the values can change with time. Another term for uniform is homogeneous.

Since the temperature is the same in the entire body, heat (entropy) must either arrive directly at every point inside or flow from point to point without the need for a temperature difference. Conduction therefore must be "superconducting", i.e. it does not lead to the production of entropy.

A uniform process can be irreversible, even if irreversibilities from the conduction of heat are excluded. We may consider the irreversibilities related to absorption and emission of radiation to be part of the body (Chapter 3.3), or we may include a viscous pressure term in a uniform model (Epilogue E.1.1).

■

Derive an expression in terms of the electrical current for the production rate of entropy in an electrical resistor operating at a steady state. Do the same for a block sliding over a surface on a viscous film. (Use the momentum current in this case.) Compare the results to Equation (2). How does Equation (2) change if you express it in terms of the current of entropy instead of the energy current associated with it?

SOLUTION:

Electric conductor: $\Pi_s = \frac{1}{T}\mathcal{D} = \frac{1}{T}\left| U_{el} I_q \right|$

Friction: $\Pi_s = \frac{1}{T}\mathcal{D} = \frac{1}{T}\left(v I_p - 0 \cdot I_p \right) = \frac{1}{T}\left| \Delta v I_p \right|$

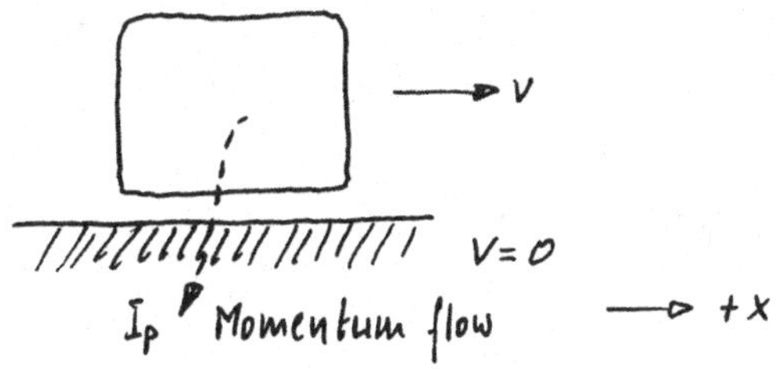

Conduction of heat: $\Pi_s = \frac{I_{E,th}\,\Delta T}{T\,T_0} = \frac{1}{T}\left| \Delta T\, I_{s,out} \right|$

The three forms are basically the same; however, since temperatures and entropy currents are not uniform, we have to be careful about the meaning of the quantities in the equations. The ambiguities are resolved in a continuum theory.

How much entropy and energy are added to 1.0 kg of silicon if the body is heated from 160 K to 640 K? (See Figure 4 for properties of silicon.)

SOLUTION: The entropy and energy exchanged are calculated in terms of the entropy capacity and the temperature coefficient of energy, respectively (Equations (13) and (19)):

$$S_e = \int_{T_1}^{T_2} K(T)\,dT \qquad Q = \int_{T_1}^{T_2} C(T)\,dT$$

$C(T) = n\bar{c}(T)$, where $\bar{c}(T)$ is presented in Figure 4. There are 35.6 moles of silicon in 1.0 kg. The entropy capacity is calculated from Equation (20):

$$\bar{c} = T\bar{k} \quad \text{with} \quad T = 640\,K \cdot T/T_D$$

Numerical integration:

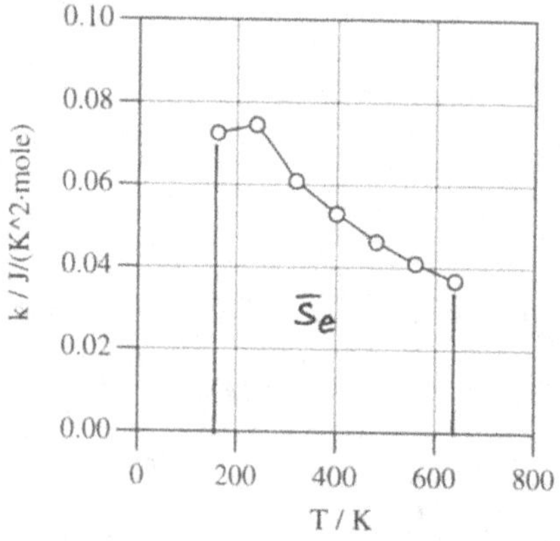

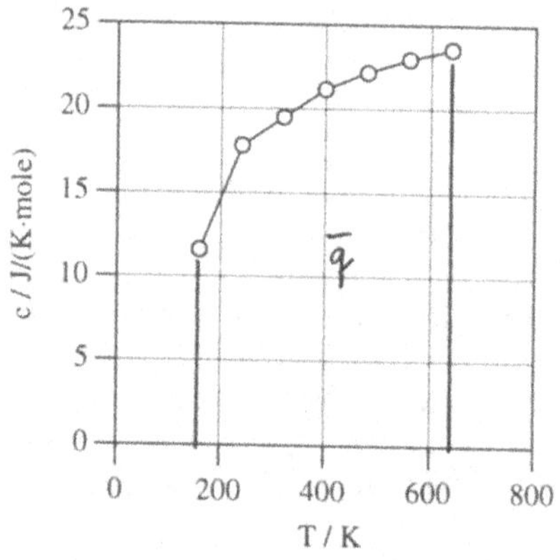

$$S_e = n \cdot \bar{s}_e = 35.6\text{ mole} \cdot 26.4\text{ J/(K·mole)} = 940\text{ J/K}$$

$$Q = n \cdot \bar{q} = 35.6\text{ mole} \cdot 9700\text{ J/mole} = 345\text{ kJ}$$

■

A 100 g piece of copper is to be heated at 30 K such that the rate of change of its temperature is 0.10 K/min. How large does the flux of entropy have to be? How large is the current of energy entering the body? (See Figure 4 for properties of copper.)

SOLUTION :

$$I_s = -K(T)\dot{T}$$

$$= -n\bar{k}(T)\dot{T} = -n\frac{1}{T}\bar{c}(T)\dot{T}$$

$$= -\frac{0.100\ \text{kg}}{0.064\ \text{kg/mole}}\cdot\frac{1}{30\text{K}}\ 1.3\frac{\text{J}}{\text{K}\cdot\text{mole}}\cdot\frac{0.10\text{K}}{60\text{ s}}$$

$$= -1.1\cdot 10^{-4}\ \text{W/K}$$

$$I_E = T I_s = 30\text{K}\left(-1.1\cdot 10^{-4}\ \text{W/K}\right) = -3.4\cdot 10^{-3}\ \text{W}$$

■

The temperature of a piece of granite lying in the sun is found to change from 20°C to 40°C in 2 hours. Its mass is 0.30 kg. At what average rate does it absorb entropy? Assume its temperature coefficient of energy to be constant over the range of temperatures considered.

SOLUTION:

$$\dot{T} = \frac{\Delta T}{\Delta t} = \frac{20\ K}{7200\ s} = 2.78 \cdot 10^{-3}\ K/s$$

$$K = \frac{1}{T} m \cdot c = \frac{1}{(273+30)K}\ 0.30\ kg\ 750\ \frac{J}{K \cdot kg}$$

$$= 0.74\ J/K^2$$

$$\Rightarrow \quad \dot{S} = K\dot{T} = 2.78 \cdot 10^{-3}\ K/s \cdot 0.74\ J/K^2$$

$$= 0.21 \cdot 10^{-3}\ W/K$$

We use the model of uniform processes:

$$\dot{S} = -I_S \quad \Rightarrow \quad I_S = -0.21 \cdot 10^{-3}\ W/K$$

■

Derive an expression for the energy density of a rigid body. (Set the energy equal to zero at room temperature.) Take its temperature coefficient of energy to be constant. What properties should the body have to achieve a high energy density?

SOLUTION:

$$E(T) = E_0 + \int_{T_0}^{T} C(T)\,dT = E_0 + C(T - T_0)$$

$$E_0(T_0) = 0 \text{ for } T_0 = 293\,\text{K}$$

$$\Rightarrow E(T) = C(T - T_0)$$

$$\rho_E = \frac{E}{V} = \frac{m \cdot c}{V}(T - T_0)$$

$$\Rightarrow \rho_E = \rho \cdot c\,(T - T_0)$$

The density and the specific temperature coefficient of energy should be large for the body to have a high energy density at a given temperature.

■

A rigid body has a constant entropy capacity in a particular range of temperatures. How much energy does it emit if its temperature drops from T_i to T_f?

SOLUTION:

Formal solution:

$$S_e = \int_{T_i}^{T_f} K(T)\, dT = K(T_f - T_i)$$

$$Q = \int_{T_i}^{T_f} C(T)\, dT = \int_{T_i}^{T_f} TK\, dT$$

$$= \frac{1}{2} K (T_f^2 - T_i^2)$$

Graphical interpretation: We may derive the results graphically using an image borrowed from hydraulics. The body is like a container for entropy, with the level representing the temperature.

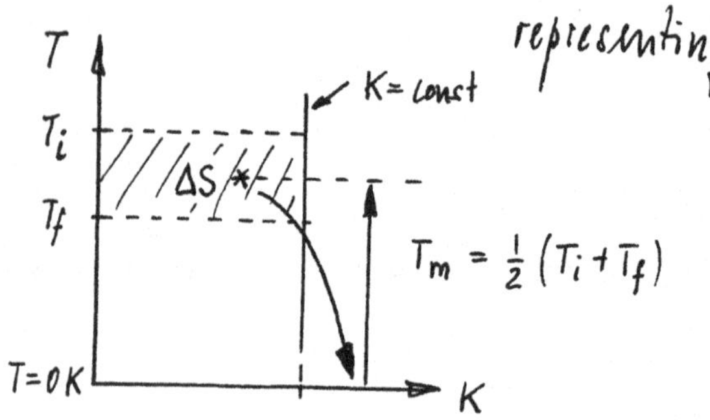

$$S_e = \Delta S = K(T_f - T_i)$$

$Q = \Delta E$: energy liberated in the fall of entropy from T_m to $T = 0\,K$:

$$Q = \left[\frac{1}{2}(T_i + T_f) - 0\right] \cdot K \cdot (T_f - T_i) = \frac{1}{2} K (T_f^2 - T_i^2)$$

■

An ideal Carnot engine is driven with the heat from 2000 liters of water at 90°C. Entropy is rejected to the environment at a temperature of 20°C. How much energy does the engine release for mechanical purposes?

SOLUTION:

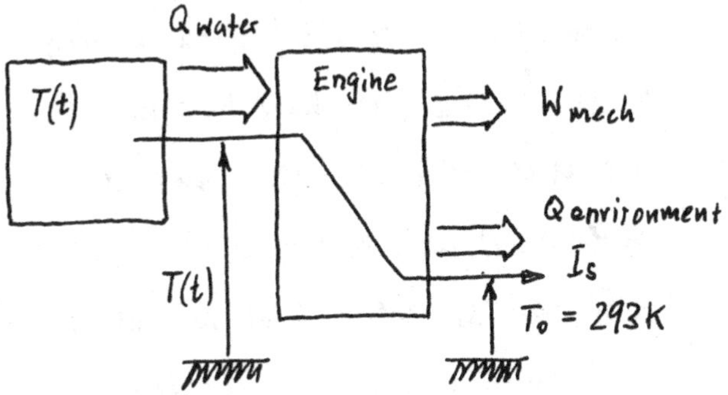

$$|W_{mech}| = |Q_{water}| - |Q_{environment}|$$

$$= C(T_i - T_f) - T_o |S_{e,water}| \qquad T_f = T_o$$

$$= C(T_i - T_f) - T_o |\Delta S_{water}|$$

$$= C(T_i - T_f) - T_f \, C \ln\left(\frac{T_i}{T_f}\right) = 60.7 \text{ MJ}$$

Graphical interpretation:

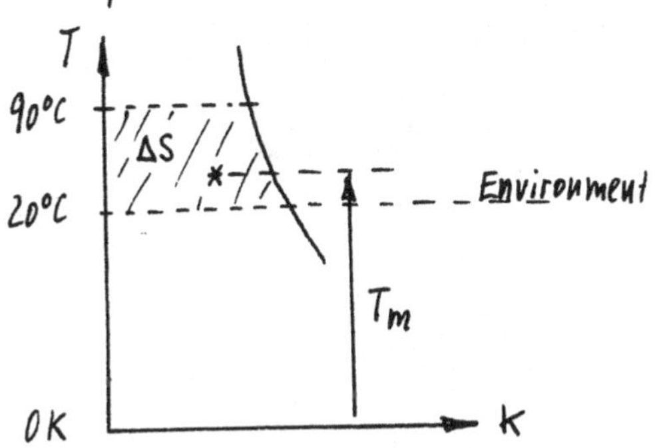

The amount of entropy ΔS drops from T_m to the level represented by the temperature of the environment.

$$|W| = (T_m - T_o)|\Delta S|$$

For the world exhibition in Seville, the architect for the Swiss pavilion suggested building a tower of ice. (The project was abandoned for environmental reasons.) Estimate the minimal amount of energy necessary to produce 1000 tons of ice in a 30°C environment.

SOLUTION:

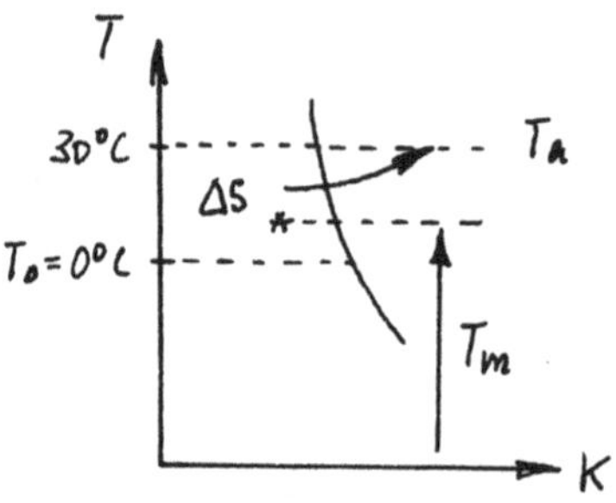

Assuming that we start with a body of water at 30°C, we have to pump the entire amount of entropy emitted by the water as it cools and freezes to the environment at 30°C. If we can use an ideal heat pump, we will need the least amount of energy for the task.

Making 0°C water (see figure):

$$|W| = (T_a - T_m)\,\Delta S \approx (T_a - T_m)\, m \cdot k(T_m)(T_a - T_0)$$
$$\approx 16\,K \cdot 10^6\,kg \cdot 14.6\,J/(K^2 kg) \cdot 30\,K$$
$$= 7.0 \cdot 10^9\,J$$

Freezing water: $|W| = (T_a - T_0)|\Delta S_{fusion}|$

$$= 30\,K \cdot 10^6\,kg \cdot 1220\,J/(K \cdot kg)$$
$$= 37 \cdot 10^9\,J$$

$$\rightarrow \quad |W_{total}| = 44 \cdot 10^9\,J$$

■

It is said that the oceanic climate is less extreme with respect to temperature variations because of the thermal buffering effect of the water, which is explained as being a result of the higher "heat capacity" of water compared to that of the land. Compare the entropy capacities per volume of water and soil or rock. If the difference appears to be too small to explain the effect upon climate, how else would you explain the phenomenon?

SOLUTION: The "specific heat capacity" (i.e., the specific temperature coefficient of energy) of water is about 4 to 5 times higher than the equivalent value for rock. However, since the density of rock is 2 to 3 times higher than that of water, the "heat capacity" per volume for water and rock is not all that much different.

Since water is transparent to some depth, large volumes of water can be heated with the Sun's light, whereas only a thin upper layer of soil or rock can store significant amounts of entropy (it takes too long for entropy to penetrate solid layers of more than just a few meters; see Chapter 3.4). Much larger volumes of water are turned over in the course of a year and release heat to the environment.

■

How does the energy content change if water undergoes an isothermal expansion at a temperature of 2°C? Is the answer different for a temperature of 20°C?

SOLUTION:

Balance of energy: $\dot{E} = - I_{E,mech} - I_{E,th}$

At 2°C:
Expansion: $I_{E,mech} > 0$
Heating: $I_{E,th} > 0$ since $\Lambda_v < 0$
$\Bigg\}\ \dot{E} < 0$

At 20°C:
Expansion: $I_{E,mech} > 0$
Heating: $I_{E,th} < 0$ since $\Lambda_v > 0$
$\Bigg\}\ \dot{E} = ?$

Unless the precise value of Λ_v is known, we cannot be sure about the sign of dE/dt.

■

A body of air of mass 20 kg at 100°C and 1.0 bar is compressed isothermally. a) How much energy has to be supplied in the mechanical process if the volume is to be reduced to 10% of its initial value? b) How large will the pressure of the air be at the end? c) If the compression is to be performed at a constant rate of change of the volume in 10 s, how large does the mechanical power have to be as a function of time?

SOLUTION: We treat air as an ideal gas subject to the theory of reversible processes developed in this chapter.

a) $$W = -\int I_{E,mech}\, dt = -\int P\dot{V}\, dt = -\int_{V_1}^{V_2} P dV$$

$$= -\int_{V_1}^{V_2} \frac{nRT}{V} dV = -nRT \ln(V_2/V_1)$$

$$\Rightarrow W = -\frac{20\text{ kg}}{0.029\text{ kg/mole}}\, 8.31 \frac{\text{J}}{\text{K}\cdot\text{mole}}\, 373\text{K} \ln(0.1) = 4.92\text{ MJ}$$

b) $T = \text{const.}$ $P_1 V_1 = P_2 V_2 \Rightarrow P_2 = (V_1/V_2) P_1 = 10 \cdot P_1 = 10^6\text{ Pa}$

c) $$I_{E,mech} = P\dot{V} = \frac{nRT}{V(t)} \dot{V}$$

$V_1 = 21.4\text{ m}^3$, $V(t) = V_1 + \dot{V}t$, $\dot{V} = (2.14 - 21.4)/10\text{ m}^3/\text{s} = -1.93\text{ m}^3/\text{s}$

$$\Rightarrow I_{E,mech} = \frac{nRT}{V_1 + \dot{V}t} \dot{V}$$

where $\dot{V} = -1.93\text{ m}^3/\text{s}$, $V_1 = 21.4\text{ m}^3$

■

Determine the entropy content of one mole of argon at 300 K and at a pressure of 1 bar. (See the values supplied in Table 5.)

TABLE 5. Values of the molar temperature coefficient of enthalpy of argon[a]

solid		liquid		gaseous	
T K	$\bar{c}_p$ $J \cdot mole^{-1}K^{-1}$	T K	$\bar{c}_p$ $J \cdot mole^{-1}K^{-1}$	T K	$\bar{c}_p$ $J \cdot mole^{-1}K^{-1}$
20	11.76	83.85	42.04	87.29	20.79
40	22.09	87.29	42.05	100	20.79
60	26.59			300	20.79
80	32.13				
83.85	33.26				

a. Values have been taken from Förstling und Kuhn (1983).

Note: In the problem statement, two important values have been omitted. The energy transferred in heating for melting one mole of argon is 1180 J; for evaporation it is 6520 J.

SOLUTION: We imagine heating 1 mole of argon at constant pressure from 0 K to 300 K. To compute the amount of entropy added, we need the

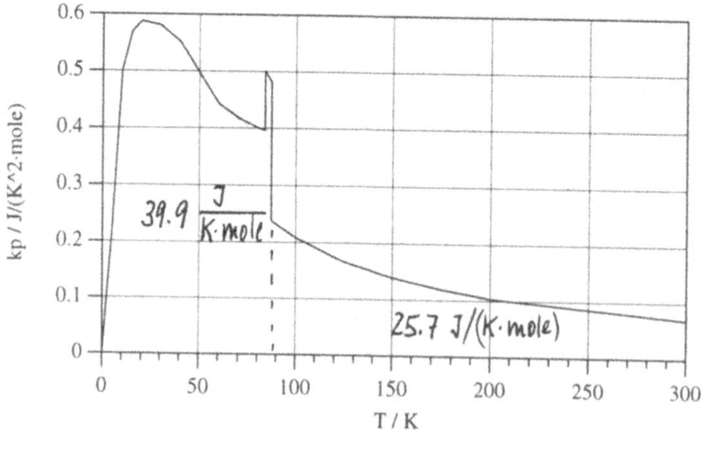

$$\bar{k}_p = \bar{c}_p / T$$

entropy capacities at constant pressure $\bar{K}_p$, and the latent entropies of fusion and of vaporization. The total entropy added is equal to the absolute entropy content, since the entropy of argon is zero at 0K.

The entropy added in heating the solid and the liquid phases is obtained by numerical integration in the $\bar{K}_p$-T diagram. The entropy added in heating the (ideal) gas at constant pressure is calculated as follows:

$$S_{e,2} = -\int I_s \, dt = \int \left(\Lambda_p \dot{P} + K_p \dot{T} \right) dt = \int K_p \dot{T} \, dt$$

$$= \int_{T_1}^{T_2} K_p(T) \, dT = n\bar{c}_p \ln\left(T_2/T_1\right)$$

$$= 1.0 \text{ mole} \cdot 20.79 \text{ J/(K·mole)} \cdot \ln\left(300/87.29\right) = 25.67 \text{ J/K}$$

Entropy added during phase transformations:

$$S_{e,3} = Q_f/T_f + Q_v/T_v = \left(\frac{1180}{83.85} + \frac{6520}{87.29} \right) \text{J/K} = 88.77 \text{ J/K}$$

Entropy content of 1 mole of argon at 300 K:

$$S(300\,\text{K}) = S_0 + S_{e,1} + S_{e,2} + S_{e,3}$$

$$= \left(0 + 39.9 + 25.67 + 88.77 \right) \text{J/K} = 154.3 \text{ J/K}$$

For more details see Försterling and Kuhn, p. 186 - 187. ■

An ideal fluid undergoes a general process represented by a curve in the *T*-*V* diagram leading from an initial point (T_i, V_i) to the final state (T_f, V_f). Show that the entropy exchanged along the path is given by

$$S_e = \int_{V_i}^{V_f} \Lambda_V dV + \int_{T_i}^{T_f} K_V dT$$

SOLUTION: The entropy exchanged is equal to the (negative) integral of the entropy current over time for the heating of the fluid (Equation 30 in Chapter 1):

$$S_e = -\int_{t_i}^{t_f} I_s \, dt$$

The entropy current, on the other hand, is given in terms of the rates of change of volume and temperature (Equation 42):

$$I_s = -\Lambda_v \dot{V} - K_v \dot{T}$$

$$\rightarrow \quad S_e = \int_{t_i}^{t_f} [\Lambda_v \dot{V} + K_v \dot{T}] dt$$

$$= \int_{t_i}^{t_f} \Lambda_v \dot{V} dt + \int_{t_i}^{t_f} K_v \dot{T} dt$$

$$= \int_{V_i}^{V_f} \Lambda_v dV + \int_{T_i}^{T_f} K_v dT$$

■

a) What type of observation shows that the latent entropy (with respect to volume) of the ideal gas must be a positive quantity? b) What must be the sign of the latent entropy with respect to pressure? What does the latter result mean?
c) Prove that the entropy capacity at constant pressure must be larger than the entropy capacity at constant volume. What is the significance of this result?

SOLUTION:

a) An ideal gas must be heated (not cooled) if its volume is to increase at constant temperature:

$$I_s = -\Lambda_V \dot{V} - K_V \dot{T} \, , \quad T = \text{const} \Rightarrow I_s = -\Lambda_V \dot{V}$$

$$\dot{V} > 0 \, , \quad I_s < 0 \text{ (heating)} \Rightarrow \Lambda_V > 0 .$$

b) Equation (47): $\Lambda_P = -\Lambda_V V/P \, ; \quad \Lambda_V > 0 \Rightarrow \Lambda_P < 0$

c) Equation (48): $K_P = \dfrac{nR}{P} \Lambda_V + K_V$

$$\Lambda_V > 0 \Rightarrow K_P > K_V$$

Consequences: An ideal gas has to be heated more at constant pressure than at constant volume for a given rise in temperature. Also, $\gamma = K_P / K_V > 1 \Rightarrow$ adiebatic sound waves are faster than isothermal ones.

■

Dry air rushes from the mountains (2500 m above sea level) into a valley (500 m above sea level). The temperature of the air in the mountains is 4°C. Before the arrival of the winds the temperature of the air in the valley is 16°C. By how much will the temperature of the air rise in the valley with the winds blowing?

SOLUTION: The dry air flowing from the mountains undergoes adiabatic compression. According to example 15, the adiabatic temperature gradient is $dT/dz|_{ad} = -0.010$ K/m.

Therefore:

$$\begin{aligned} T(500\text{ m}) &= T(2500\text{ m}) + dT/dz|_{ad} \cdot \Delta z \\ &= 4°C + (-0.010\text{ K/m}) \cdot (-2000\text{ m}) \\ &= 24°C \end{aligned}$$

The temperature in the valley increases by 8°C because of the dry wind.

■

Derive the expression for the pressure as a function of height in the earth's atmosphere if pressure and volume are related by the adiabatic condition for dry air.

SOLUTION :

Hydrostatic equilibrium: $\dfrac{dP}{dz} = -\rho g$

Adiabatic relation: $PV^{\gamma} = P_0 V_0^{\gamma}$

$$V = m/\rho \Rightarrow P/\rho^{\gamma} = P_0/\rho_0^{\gamma}$$

$$\Rightarrow \rho = \frac{\rho_0}{P_0^{1/\gamma}} P^{1/\gamma}$$

$$\Rightarrow \frac{dP}{dz} = -g \frac{\rho_0}{P_0^{1/\gamma}} P^{1/\gamma}$$

Solution of DE: $\displaystyle\int_{P_0}^{P} \frac{dP}{P^{1/\gamma}} = -g \frac{\rho_0}{P_0^{1/\gamma}} \int_0^z dz$

$$\frac{1}{1-1/\gamma}\left(P^{1-1/\gamma} - P_0^{1-1/\gamma}\right) = -\frac{g\rho_0}{P_0^{1/\gamma}} z$$

$$P^{1-1/\gamma} = P_0^{1-1/\gamma} - \left(1-\frac{1}{\gamma}\right)\frac{g\rho_0}{P_0^{1/\gamma}} z$$

■

Why does the entropy content of the ideal gas remain constant as a result of an adiabatic process? Determine the special forms of the equations of balance of entropy and energy for such a process. What happens to the energy of the ideal gas during adiabatic expansion?

SOLUTION: For an adiabatic process the heating is zero: $I_s = 0$. If a gas conforms to the model discussed in Section 2.3, its processes must be reversible: $\Pi_s = 0$. From the balance of entropy

$$\dot{S} + I_s = \Pi_s$$

we conclude that $\dot{S} = 0$, and the process is isentropic (constant entropy).

The laws of balance of entropy and energy are:

$$\dot{S} = 0$$
$$\dot{E} = -I_{W,mech}$$

(since $I_W = T I_s = 0$). During adiabatic expansion, the energy of the ideal gas decreases since $I_{W,mech} = P\dot{V}$:

$$\dot{E} = -P\dot{V}\ , \quad \dot{V} > 0 \Rightarrow \dot{E} < 0$$

■

According to the result obtained for adiabatic processes of the ideal gas, the ratio of the latent entropy and the entropy capacity must be proportional to *T/V*. As a result, does the entropy capacity depend on *T* or *V*? How do C_V and C_P depend on temperature or volume?

SOLUTION: First we demonstrate that $T/V \sim \Lambda_V / K_V$

Adiabatic process (Equ. 56): $$\frac{dT}{dV} = -\frac{\Lambda_V}{K_V}$$

or Equation (61): $$\frac{\dot{P}}{P} + \gamma \frac{\dot{V}}{V} = 0$$

Equation of state: $$\dot{P} V + \dot{V} P = nR\dot{T}$$

$$\Rightarrow \quad \frac{\dot{T}}{T} + (\gamma - 1)\frac{\dot{V}}{V} = 0$$

$$\Rightarrow \quad \frac{dT}{dV} + (\gamma - 1)\frac{T}{V} = 0 \quad \Rightarrow \quad \frac{\Lambda_V}{K_V} = (\gamma - 1)\frac{T}{V} \sim \frac{T}{V}$$

Entropy capacity: $\Lambda_V = nR/V \quad \Rightarrow \quad K_V \sim 1/T$

$$\Rightarrow \quad C_V = T K_V = \text{const.}$$
$$C_P = \gamma C_V = \text{const.}$$

■

Derive the equations of adiabatic change of the ideal gas, using the result of Example 17, i.e., the equation which determines the change of the entropy of the ideal gas.

SOLUTION:

Example 17: $$\Delta S = nR \ln\left(\frac{V_f}{V_i}\right) + n\bar{c}_v \ln\left(\frac{T_f}{T_i}\right)$$

Adiabatic reversible process: $\Delta S = 0$

$$\Rightarrow \quad 0 = \frac{R}{\bar{c}_v} \ln\left(\frac{V_f}{V_i}\right) + \ln\left(\frac{T_f}{T_i}\right)$$

$$R/\bar{c}_v = \gamma - 1$$

$$\Rightarrow \quad \ln\left[\left(\frac{V_f}{V_i}\right)^{\gamma-1} \cdot \frac{T_f}{T_i}\right] = 0$$

$$\Rightarrow \quad V_i^{\gamma-1} T_i = V_f^{\gamma-1} T_f$$ as in Equation (65)

■

From what you know about the properties of water how would an adiabat look in the *T-V* diagram in the range 0 – 4°C? Try to sketch an adiabat for water for a process in which the temperature changes from 2°C to 6°C.

SOLUTION:

Equation of an adiabat: $\frac{dT}{dV} = -\frac{\Lambda_V}{K_V}$ (Equation 56)

For water: a) $0°C < T < 4°C$: $\Lambda_V < 0 \Rightarrow dT/dV > 0$

b) $4°C < T < 100°C$: $\Lambda_V > 0 \rightarrow dT/dV < 0$

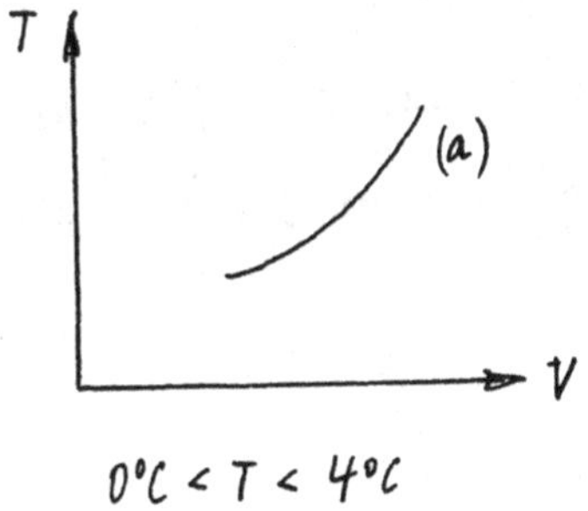

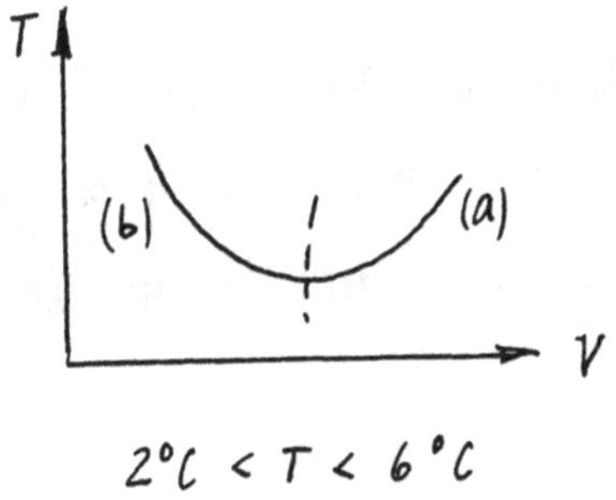

■

Draw the curve for an isobaric process of air in the *T-S* diagram. Repeat the problem for the same body of air for a process at a higher value of the pressure.

SOLUTION :

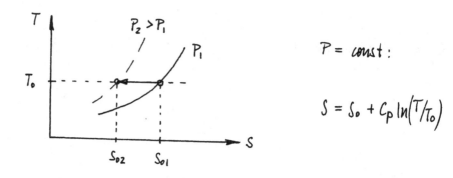

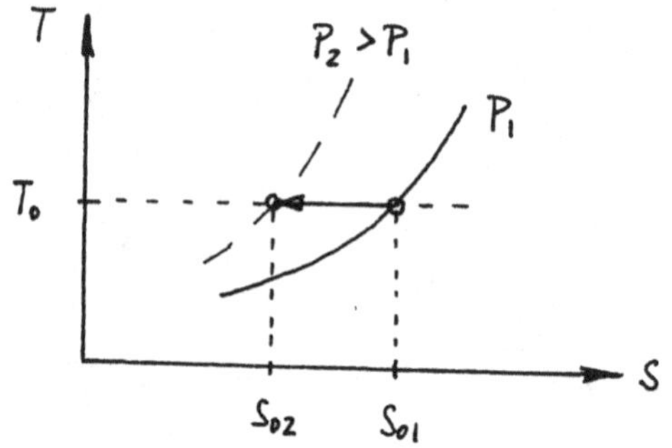

$P = const:$

$$S = S_0 + C_p \ln(T/T_0)$$

For a given value of the temperature, the volume of the gas is smaller at higher pressure. Smaller volume at equal temperature means that the entropy content must be smaller.

■

A body of air is heated at constant pressure. What fraction of the entropy added remains in the body? What fraction of the energy added as a result of heating remains there?

SOLUTION :

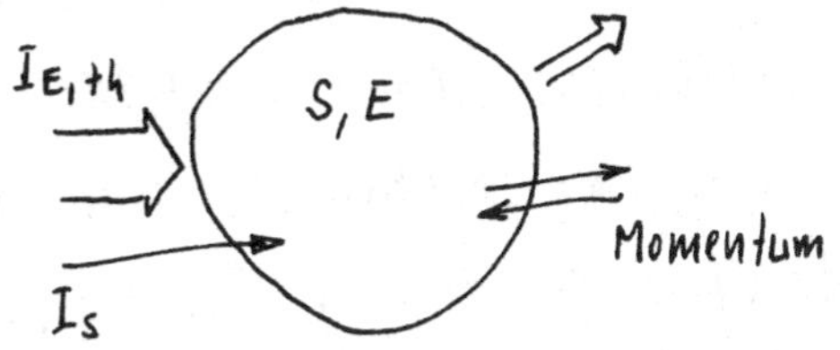

Since the body is only heated and not cooled, all the entropy supplied to it also remains inside.

For energy, on the other hand, the situation is different, since part of the energy added together with entropy leaves the body as a result of expansion :

$$\Delta E = Q + W \quad , \quad W = -P\Delta V$$

$$PV = nRT \quad \rightarrow \quad P\Delta V = nR\,\Delta T$$

$$Q = n\bar{c}_p\,\Delta T$$

$$\Rightarrow \quad \Delta E = Q - nR\frac{Q}{n\bar{c}_p}$$

$$\Rightarrow \quad \frac{\Delta E}{Q} = 1 - \frac{R}{\bar{c}_p} = 1 - \frac{\bar{c}_p - \bar{c}_v}{\bar{c}_p} = 1 - (1 - 1/\gamma)$$

$$\Rightarrow \quad \frac{\Delta E}{Q} = 1/\gamma$$

■

Air having a mass of 5 g, at a pressure of 38 bar and a temperature of 650°C, is heated inside a cylinder by burning some injected fuel. The amount of energy added by the burning fuel is 7.5 kJ. The piston moves in such a way as to leave the pressure of the air constant. (This corresponds to a step in the Diesel process.) Assume that the fuel added does not change the properties of the air in the cylinder. a) How much energy is exchanged as a result of the change of volume of the air? b) Calculate the change of the energy of the gas.

SOLUTION:

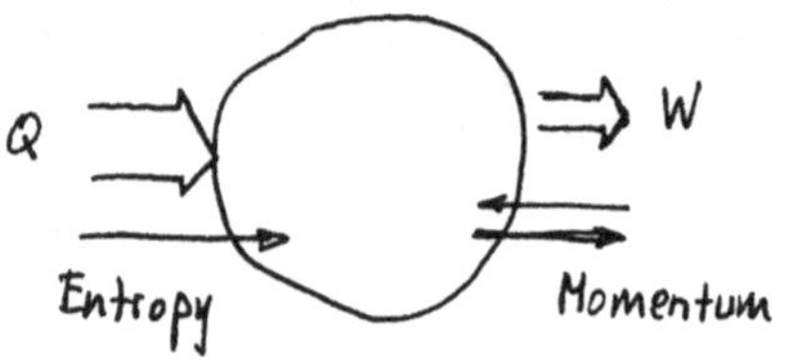

MODEL: We assume the body of air to be heated from the outside. It expands at constant pressure. We therefore are dealing with two processes and two modes of energy exchange.

a) $W = -P\Delta V$, $P = const$

$Q = m c_p \Delta T$, $Q = 7500\ J$

$$P\Delta V = \frac{mR}{M_0}\Delta T \quad \Rightarrow$$

$$\Rightarrow \quad W = -\frac{mR}{M_0}\frac{Q}{m c_p} = -\frac{RQ}{M_0 c_p}$$

$$W = -\frac{8.31 \cdot 7500}{0.029 \cdot 1006}\ J = -2.14\ kJ$$

b) Balance of energy: $\Delta E = Q + W$

$= 7500\ J + (-2140\ J)$

$= 5360\ J$ ■

A bubble of air with an initial diameter of 5.0 mm starts rising from the bottom of a pond at a depth of 5 m. The temperature of the water is 6°C at the bottom and 15°C at the surface. Assume the bubble to have the same temperature as the surrounding water at all times. Neglect the effects of surface tension. a) Calculate the radius of the bubble shortly before it reaches the surface. b) Approximately estimate the amount of energy exchanged as a consequence of heating while the bubble is rising.

SOLUTION: We model the process as that of the ideal gas being heated by the surrounding water as the pressure changes.

a) $$\left.\begin{aligned} P_1 V_1 &= nRT_1 \\ P_2 V_2 &= nRT_2 \end{aligned}\right\} \Rightarrow \frac{V_2}{V_1} = \frac{T_2 \cdot P_1}{T_1 \cdot P_2} \quad ; \quad \frac{V_2}{V_1} = \left(\frac{r_2}{r_1}\right)^3$$

Assume $P_2 = 10^5$ Pa ; $P_1 = P_2 + \rho g h$

$$\frac{r_2}{r_1} = \sqrt[3]{\frac{288\,(10^5 + 1000 \cdot 10 \cdot 5)}{10^5 \cdot 279}} = 1.16 \Rightarrow d_2 = 5.8\text{ mm}$$

b) Balance of energy: $Q = \Delta E - W$, $W \approx -\bar{P}\Delta V$

$$\Delta E = m c_v \Delta T$$

$$P_1 V_1 = \frac{mR}{M_0} T_1 \Rightarrow m = 1.23 \cdot 10^{-7}\text{ kg}$$

$$Q = m c_v \Delta T + \bar{P}\Delta V$$
$$= 1.23 \cdot 10^{-7} \cdot 717 \cdot 9\text{ J} + 1.25 \cdot 10^5 \frac{4}{3}\pi \left(0.0058^3 - 0.0050^3\right)\text{J}$$
$$= 7.92 \cdot 10^{-4}\text{ J} + 3.67 \cdot 10^{-2}\text{ J}$$
$$= 3.75 \cdot 10^{-2}\text{ J}$$

Question: Can the bubble be heated fast enough for its temperature to be equal to that of the water at all times? ∎

Use the law of hydrostatic equilibrium for a column of gas extending from the center of the Sun to its surface to estimate the pressure at the center. The gas at the center of the Sun is ideal. Determine the temperature at the center from a rough estimate of the density. How large is the contribution of radiation to the pressure at the center of the sun?

SOLUTION:

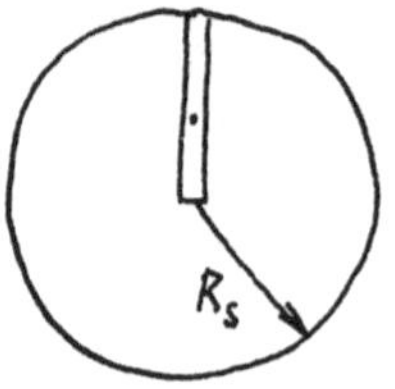

Take a column with the mean density of the Sun which is 1400 kg/m³ (mass: $2 \cdot 10^{30}$ kg, radius $R_s = 700\,000$ km). For the gravity we take the value estimated at half the radius, assuming that half the mass is below that point:

$$\bar{g} = \frac{G\,M/2}{(R_s/2)^2}$$

$$\Rightarrow \Delta P = -\bar{\rho}\,\bar{g}\,R_s = -\bar{\rho}\,\frac{2GM}{R_s} = -5.3 \cdot 10^{14}\ \text{Pa}$$

$$\rightarrow P_c \approx 5.3 \cdot 10^{14}\ \text{Pa}$$

Estimate of the central temperature:

$$P = \frac{R}{M_0}\rho T$$

with $M_0 = 0.0005$ kg/mole (half of that of hydrogen, since there are two particles per hydrogen atom).

$$T_c = \frac{M_0 P_c}{R\,\bar{\rho}} = 2 \cdot 10^7\ \text{K}$$

(a better value is around $15 \cdot 10^6$ K)

Radiation pressure: $P_R = \frac{1}{3}\rho_E = \frac{1}{3} a T^4 = \frac{1}{3} 7.56 \cdot 10^{-16} (2 \cdot 10^7)^4$ Pa

$$\approx 4 \cdot 10^{13}\ \text{Pa} \approx 0.075\, P_c$$

The estimate is too high. ■

Consider a paddle wheel inside a tank containing some viscous fluid. As the fluid is stirred, entropy is produced. If the tank is insulated, the temperature of the systems must rise. Assume the Gibbs fundamental form to hold for the system and derive the relation between the energy dissipated and the entropy produced. Do the same for an immersion heater that is heating up.

SOLUTION:

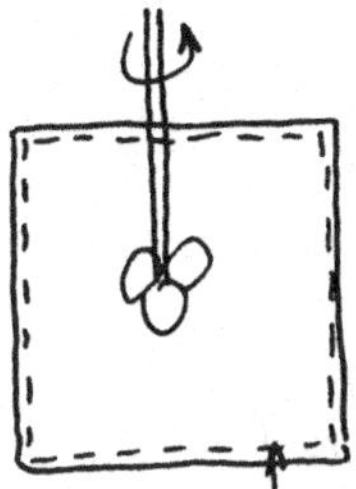

Gibbs fundamental form:

$$\dot{E} = T\dot{S}$$

Balance of energy:

$$\dot{E} = -I_{E,mech}$$

Dissipation:

$$\mathcal{D} = |I_{E,mech}|$$

$$\Rightarrow \mathcal{D} = T\dot{S}$$

The derivation is analogous for the second case, only with $I_{E,mech}$ replaced by $I_{E,el.}$.

■

Suppose we raise the temperature of an amount of water from T_L to T_H by stirring, and then drive an ideal Carnot engine with the entropy of the water released to the environment at temperature T_L. a) Calculate the amount of energy dissipated. b) How large is the amount of energy gained for the mechanical process? c) How large is the loss of availability? d) Why is the loss different from how it is defined in Equation (46) of Chapter 1?

SOLUTION:

a) $\Delta E = W$, all of the energy added (W) is dissipated

$$\Delta E = C\Delta T = C(T_H - T_L) \rightarrow W_{diss} = C(T_H - T_L)$$

b)

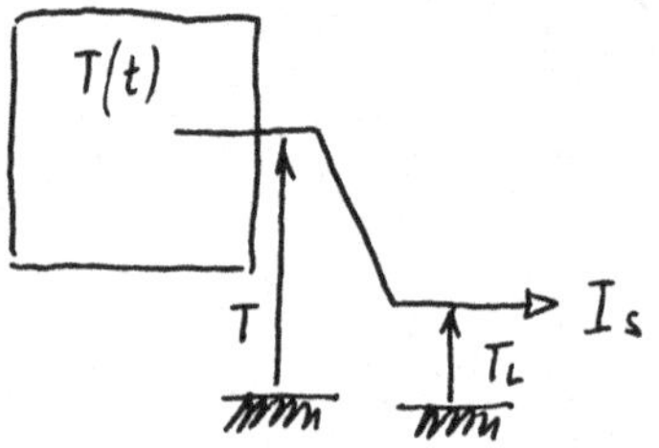

$$W = |\Delta E| - T_L |S_e|$$

$$|S_e| = C \cdot \ln(T_H/T_L)$$

c) If we only consider the second step of driving the ideal Carnot heat engine, there is no loss of availability.

d) In the first step (stirring), an amount of entropy equal to $S_{prod} = C \ln(T_H/T_L)$ was produced. Therefore, according to Equation (46) of Chapter 1 we have

$$W_{loss} = T_L C \ln(T_H/T_L).$$

Actually, if we interpret the second term on the right-hand side of $W = |\Delta E| - T_L|S_e|$ as a loss, we do get the same answer as in Equation (46).

■

Take a fluid, such as the ideal gas, which may undergo heating and cooling at constant temperatures only (Carnot processes). a) Express the work done by such a fluid in an operation which does not necessarily have to be a closed cycle, in terms of the changes of entropy and energy, and the entropy absorbed. b) Prove that the motive power of a Carnot engine is a special case of this work estimate. c) Calculate the work done in a process in which no entropy is exchanged. d) What is the work done by a fluid in an isothermal process in which entropy is absorbed but not emitted? e) Assume that dissipation cannot increase the energy emitted by the fluid as a consequence of the mechanical process. Show that this assumption leads to the result that the change of the entropy content of the fluid is larger than the entropy exchanged. (Entropy must have been produced.)

SOLUTION: a) We can start from the expression combining the laws of balance of entropy and of energy for Carnot processes as in Eq.(2) of Example 37. We have to set the rate of production of entropy equal to zero since ideal fluids can undergo reversible operations only. The expression has to be integrated over time. Note that heating and cooling take place during distinct intervals of time which is necessary if the temperature of the fluid is to be unique at any given moment:

$$\int_{t_i}^{t_f} \dot{E}dt + \int_{t_i}^{t_f} I_{E,mech}\,dt + \int_{t_i}^{t_f} (T - T_o)I_{s,in}\,dt - \int_{t_i}^{t_f} T_o\dot{S}dt = 0 \qquad (1)$$

The first term equals the change of the energy of the fluid, while the second one is equal to the negative energy exchanged due to change of volume. Since the temperatures involved in heating are constant, the third term is equal to minus the product of the difference of temperatures and the entropy absorbed S^+ (remember that a flux is negative for a current entering a body). The integral of the rate of change of the entropy is the change of the entropy content. In summary, the mechanical energy exchanged by the body is given by:

$$W = \Delta E - (T - T_o)S^+ - T_o\Delta S \qquad (2)$$

b) A Carnot cycle is a process which returns the engine, i.e. its driving agent, to the same state at the end of each complete cycle of operation. We can therefore assume the engine to be in the same condition again after a complete cycle. In particular, the energy of the engine must be the same again which means that the change of energy in (2) vanishes. The same holds for the change of the entropy of the engine; remember that a Carnot engine works by absorbing and emitting entropy. In one cycle it must disengage all the entropy it absorbs. Therefore, with $\Delta E = 0$ and $\Delta S = 0$, the work estimate (2) becomes:

$$W = -(T - T_o)S^+ \qquad (3)$$

which is equivalent to Equation (11) of Chapter 1.

c) If entropy is not exchanged in a process, S^+ becomes zero. Also, since entropy is conserved in the operations considered, the change of the entropy content vanishes. Therefore:

$$W = \Delta E \qquad (4)$$

which is a special case of the energy balance in thermomechanical processes as expressed by Equation (120). You only have to set $Q = 0$ in that equation. The process described is an adiabatic change.

d) In (2), we replace the change of the entropy content by the difference of the entropy absorbed and the entropy emitted. Since the entropy emitted is zero we find that

$$W = \Delta E - T \cdot S^+ \qquad (5)$$

It is found that for air and other gases at normal conditions the change of the energy in isothermal processes is almost zero. In this case the energy emitted in expansion is equal to the energy absorbed in heating.

e) If we allow for dissipation, the work estimate (2) changes into an inequality. The energy W emitted by an engine as a consequence of working is smaller in magnitude in the presence of dissipation. Taking into account the sign of W we obtain:

$$W \geq \Delta E - (T - T_o)S^+ - T_o \Delta S \qquad (6a)$$

or

$$T_o \Delta S \geq \Delta E - W - (T - T_o)S^+ \qquad (6b)$$

The first two terms on the right hand side add up to the energy exchanged as a consequence of heating. For Carnot processes this expression turns into:

$$Q = TS^+ - T_o S^- \qquad (7)$$

Introducing this term into (6b) leads to the following equation of balance of entropy:

$$\Delta S - (S^+ - S^-) \geq 0 \qquad (8)$$

This is equivalent to

$$\Delta S + \int_{t_i}^{t_f} I_s \, dt \geq 0 \qquad (9)$$

The equation expresses the balance of a quantity which is produced in dissipative processes.

■

Consider the following strongly simplified model of a gas of noninteracting point particles. N particles are contained in a cube of side L. Assume one-third of the particles to travel in each of the three directions parallel to the sides of the cube. All particles have the same speed. a) Show that the pressure of the particle gas is given by

$$P = \frac{1}{3}\frac{N}{V} v p$$

where v and p are the speed and the momentum of a single particle, respectively, and V is the volume of the cube. b) Apply this result to an ideal gas of material particles. Derive the relation between the pressure and the energy density of the gas. c) Again for the material ideal gas, derive the expression for the temperature coefficient of energy. Which gases have such a value for the coefficient? d) Apply the idea to the photon gas. Show that the relation between pressure and energy density is given by Equation (88).

SOLUTION:

a)

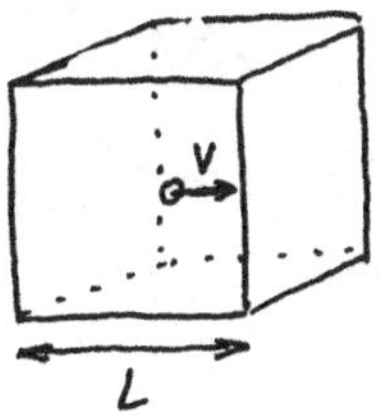

The pressure is equal to the momentum flux density of momentum transferred to a wall:

$$P = I_p/A = \frac{1}{A}\left(N/3 \cdot p_e/\Delta t\right) \quad \text{where } p_e = 2p.$$

Δt is the time between consecutive collisions of the particle with the same wall: $\Delta t = 2L/v$:

$$P = \frac{1}{3} N \frac{2pv}{L^2\, 2L} = \frac{1}{3}\frac{N}{V} p \cdot v$$

b) $\rho_E = E/V = N\frac{1}{2}mv^2$; $P = \frac{1}{3} N/V\, mv \cdot v \Rightarrow P = \frac{2}{3}\rho_E$

c) $P = nRT/V \rightarrow nRT/V = \frac{2}{3}\rho_E$; $C_v = \partial E/\partial T = \frac{3}{2}nR$ (monatomic gas)

d) $P = \frac{1}{3} N/V\, p \cdot c = \frac{1}{3} N/V\, \bar{e} = \frac{1}{3}\rho_E$ ($\bar{e}$: energy per photon)

Can you explain the difference between thermostatics and thermodynamics? Use Callen's definition of the task of thermostatics (which he calls thermodynamics) found in Footnote 16 for a starting point.

SOLUTION: Statics is concerned with the computation of equilibrium states. As Example 41 demonstrates, this may be accomplished by employing a variational principle: a function of the independent variables of the system becomes maximal or minimal in equilibrium. Similar approaches are known from statics in mechanics.

Dynamics, on the other hand, requires "equations of motion", i.e., the law of balance of entropy ($\dot{S} + I_S = \Pi_S$) in thermal physics, and proper constitutive laws for the particular processes investigated.

■

INTERLUDE

Solutions of Selected Problems

Derive the differential equation for the path which represents an isobaric process in the *T-V* diagram. Calculate the isobars for the ideal gas.

SOLUTION:

The differential equation for an isobar follows directly from Equation (5). We only have to set the time rate of change of the pressure equal to zero:

$$\frac{\partial p}{\partial V}\dot{V} + \frac{\partial p}{\partial T}\dot{T} = 0$$

Since $\partial p/\partial V < 0$ always, and since $dT/dt \neq 0$ during such a process, we can change this equation into

$$\frac{dV}{dT} = -\frac{\partial p/\partial T}{\partial p/\partial V}$$

The partial derivatives of the pressure of the ideal gas are $\partial p/\partial V = -P/V$ and $\partial p/\partial T = P/T$. Therefore the differential equation for the isobars is $dV/dT = V/T$ which has the following solution: $T \sim V$. The isobars of the ideal gas are straight lines going through the origin of the *T-V* quadrant. Naturally, we could have obtained this result by simply setting $P =$ const. in the equation of state of the ideal gas.

■

A capacitor made up of two (large) parallel plates with variable separation is a simple electromechanical device. Choose independent variables appropriate to the problem and write the "equation of state" for this system analogous to a thermal equation of state. Calculate the curves which correspond to isobars in the thermal case.

SOLUTION:

We choose as independent variables an electrical and a mechanical variable. As in the thermal case, we employ as the electrical quantity the intensive one which here is the voltage U. The separation x of the plates can serve as the extensive mechanical quantity analogous to the volume of a fluid. There is a quantity comparable to the pressure of a fluid, namely the force F acting between the two plates divided by the surface area A of the plates (i.e. the momentum current density in the electromagnetic field). Since the force is attractive we have a case of tension rather than compression. Therefore I shall introduce the tension $\mathcal{T} = F/A$ which is a negative quantity, i.e. $\mathcal{T} < 0$ always. From electromagnetic theory we know that

$$\mathcal{T} = -\frac{1}{2}\varepsilon_o \frac{U^2}{x^2}$$

where ε_o is the permittivity constant. This equation can serve as the equivalent of an equation of state. The rate of change of the tension is calculated in terms of the two independent variables:

$$\dot{\mathcal{T}} = \frac{\partial \mathcal{T}}{\partial x}\dot{x} + \frac{\partial \mathcal{T}}{\partial U}\dot{U}$$

with the partial derivatives

$$\frac{\partial \mathcal{T}}{\partial x} = -2\frac{\mathcal{T}}{x} \quad , \quad \frac{\partial \mathcal{T}}{\partial U} = 2\frac{\mathcal{T}}{U}$$

The differential equation for the "isobars" follows from the equation given above by setting the time derivative of $\mathcal{T}$ equal to 0:

$$\frac{dx}{dU} = -\frac{\partial \mathcal{T}/\partial U}{\partial \mathcal{T}/\partial x} = \frac{x}{U}$$

Introducing the partial derivatives for our system we obtain $x \sim U$ as the solution for "isobars". These curves therefore are straight lines going through the origin of the U-x quadrant.

■

Check whether the constitutive inequalities (3), (11), and (12), and Equation (13) hold for the ideal gas.

SOLUTION:

The partial derivatives of the functions are:

$$\partial p/\partial V = -nRT/V^2 = -P/V < 0$$
$$\partial p/\partial T = nR/V = P/T > 0$$
$$\partial V/\partial T = nR/P = V/T > 0$$
$$\partial p^*/\partial \rho = P/\rho > 0$$

This means that the volume of the ideal gas will increase if its temperature is increased at constant pressure; also, the gas satisfies the condition of Equation (11) so that sound may propagate in it.

■

Liquids have very small compressibilities. Therefore, their pressures must increase considerably if their temperatures are increased at constant volume. Calculate the increase of pressure per 1 K for water at 20°C, and give an approximate formula for the change of pressure with temperature.

SOLUTION:

The isothermal compressibility of water is equal to $4.58 \cdot 10^{-10}$ m^{2}N^{-1} at 20°C and 1 bar (see Table A.1), while the coefficient of thermal expansion of water at 20°C is equal to $2.07 \cdot 10^{-4}$ K^{-1} (see Table A.2). According to Equation (19), the pressure coefficient of water at 20°C turns out to be

$$\beta = \frac{\alpha_V}{P\kappa_T} = \frac{2.07 \cdot 10^{-4}}{1 \cdot 10^5 \cdot 4.58 \cdot 10^{-10}} \mathrm{K}^{-1} = 4.52\,\mathrm{K}^{-1}$$

Therefore

$$\Delta P = P\beta\Delta T = 4.52 \cdot 10^5\ \mathrm{Pa/K}\,\Delta T$$

which holds approximately near a temperature of 20°C.

■

For the parallel plate capacitor, introduce a quantity which is equivalent to the isothermal compressibility of a fluid. What does *isothermal* mean in this context?

SOLUTION:

The compressibility is defined as the derivative of the volume with respect to the pressure (Equation (18)). the equivalent quantities are separation x and tension $\mathcal{T}$ while the voltage takes the role of the temperature (isothermal therefore means constant voltage). Therefore we should have:

$$\kappa_U = -\frac{1}{x}\frac{\partial x(\mathcal{T},U)}{\partial \mathcal{T}}$$

Since $\partial x/\partial \mathcal{T} = (\partial \mathcal{T}/\partial x)^{-1}$, we can calculate the "isovoltaic compressibility":

$$\kappa_U = -\frac{1}{x}\left(-2\frac{\mathcal{T}}{x}\right)^{-1} = \frac{1}{2\mathcal{T}}$$

The tension is a negative quantity which means that $\kappa < 0$. High κ means that it is simple to stretch the capacitor ("soft" capacitor). Since we have a state of tension rather than of compression we could call this quantity the tensile strength of the system. It is similar to the strength of a rubber band.

■

For the parallel plate capacitor with variable separation, introduce and derive quantities which are analogous to the latent heat and the heat capacity. Calculate the equation which holds in processes for which no charge is exchanged. These processes are comparable to those of a gas without the exchange of heat (adiabatic processes).

SOLUTION:

Since charge is a conserved quantity we can introduce the charge capacity and the "latent charge" as the partial derivatives of the charge function of the capacitor. With

$$q(x,U) = \varepsilon_o A \frac{1}{x} U$$

we get

$$\Lambda_x = \frac{\partial q}{\partial x} = -\varepsilon_o A \frac{U}{x^2}$$

$$K_x = \frac{\partial q}{\partial U} = \varepsilon_o A \frac{1}{x}$$

The former we call the "latent charge", while the latter represents the commonly used charge capacity. The "charging" of such a system therefore obeys the relationship

$$I_q = -\Lambda_x \dot{x} - K_x \dot{U} = \varepsilon_o A \frac{U}{x^2} \dot{x} - \varepsilon_o A \frac{U}{x} \dot{U}$$

where I_q is the electrical current. If we electrically isolate the system, processes satisfy

$$I_q = 0$$

which leads to the differential equation for such "adiabatic" processes:

$$\frac{dx}{dU} = \frac{x}{U}$$

Its solution is $x \sim U$ which means that the "adiabats" are straight lines which go through the center of the U-x diagram.

■

We can define a general compressibility as follows:

$$\kappa = -\frac{1}{V}\frac{dV}{dP}$$

Demonstrate that the isothermal and the adiabatic compressibilities follow from this equation by specializing dV/dP to either isothermal or adiabatic processes.

SOLUTION:

The derivative dV/dP is equal to

$$\dot{V}/\dot{P} = \frac{1}{\dot{P}}\left[\frac{\partial V}{\partial P}\dot{P} + \frac{\partial V}{\partial T}\dot{T}\right] = \frac{\partial V}{\partial P} + \frac{\partial V}{\partial T}\frac{dT}{dP}$$

a) Isothermal processes: in this case $dT/dP = 0$ and the general compressibility reduces to the isothermal one.
b) According to Equation (39) we have $dT/dP = -\Lambda_P/K_P$ on an adiabat. Therefore

$$\begin{aligned}\frac{\partial V}{\partial T}\frac{dT}{dP} &= \left[\frac{\partial p}{\partial V}\right]^{-1}\frac{\partial p}{\partial T}\frac{\Lambda_P}{K_P} \\ &= \frac{1}{K_P}\left[\frac{\partial p}{\partial V}\right]^{-1}\frac{\partial p}{\partial T}\left[\frac{\partial p}{\partial V}\right]^{-1}\Lambda_V \\ &= -\frac{K_p - K_V}{K_P}\left[\frac{\partial p}{\partial V}\right]^{-1} \\ &= -\frac{K_p - K_V}{K_P}\frac{\partial V}{\partial P}\end{aligned}$$

which is a consequence of Equations (7b), (40a), (40b), and (7a). From this results

$$\frac{dV}{dP} = \frac{1}{\gamma}\frac{\partial V}{\partial P}$$

which leads to the adiabatic compressibility.

■

According to the derivation in the Prologue, the momentum inductance per length of a column of fluid is given by

$$L_p^* = \left[A\rho \frac{dP}{d\rho} \right]^{-1}$$

Derive the relationship between the inductance and the compressibility.

SOLUTION:

According to Equations (9) and (10) we have

$$\dot{V} = -V\,\dot{\rho}/\rho$$

Therefore

$$\frac{dP}{d\rho} = -\frac{V}{\rho}\frac{dP}{dV} = \frac{1}{\rho\kappa}$$

If we introduce this in the equation given above we find that

$$L_p^* = \kappa/A$$

Here, A is the surface area of the column of fluid.

■

The speed of sound in air has been measured for a constant pressure of 101.3 kPa, and for several temperatures (Table 1). Determine whether the square of the speed of sound is proportional to the temperature of the air, and calculate the value of the ratio of the specific heats. Would the data support the assumption that the speed of sound is proportional to the temperature?

TABLE 1. Speed of sound in air for different temperatures

T / K	c / ms^{-1}	c^2	γ
233	307	94249	1.41
253	319	101760	1.40
273	332	110220	1.41
293	344	118340	1.41
313	355	126030	1.40

SOLUTION:

The thermal equation of state (2) expressed in terms of the mass density ρ is

$$P = \frac{R}{M_o}\rho T$$

where M_o is the molar mass of the gas. We therefore get

$$c^2 = \gamma\frac{P}{\rho} = \frac{R}{M_o}T$$

or

$$\gamma = c^2\frac{M_o}{RT}$$

Both the speed of sound and its square nicely fit linear relationships. However, the speed is not proportional to the temperature T (it is proportional to a linear function of T), while the square of the speed is. The ratio of the specific heats turns out to be 1.40.

■

Derive the relationship between the speed of propagation of sound in a fluid and its compressibility. The speed of sound in water at a temperature of 20°C is 1483 m/s. What is the adiabatic compressibility of water?

SOLUTION:

It is clear that the propagation of sound through a fluid has something to do with the compressibility of this fluid. It appears reasonable that the speed of sound increases with decreasing compressibility. In a perfectly rigid body (compressibility equal to zero) signals would travel infinitely fast. Here we can demonstrate this relationship. With the help of Equation (7b) and the definition of the adiabatic compressibility (see Equations (57) and (58)) we get:

$$c^2 = \gamma \frac{\partial p^*}{\partial \rho} = \gamma \left(-\frac{V}{\rho} \frac{\partial p}{\partial V} \right) = \frac{1}{\rho \kappa_s}$$

Isothermal and adiabatic compressibility are almost the same. According to the previous example we get:

$$\kappa_s = \frac{1}{\rho c^2} = \frac{1}{998.2 \cdot 1483^2} \text{Pa}^{-1} = 0.46 \cdot 10^{-9} \text{Pa}^{-1}$$

■

A Carnot-type cycle is not possible only in thermodynamics. We can let a parallel plate capacitor undergo a cycle which has all the features of the process proposed by Carnot.[1] Consider a capacitor made of two large parallel plates which are separated by a small distance. The surface area of the capacitor plates is assumed to be 1 m^2. Here, voltage, separation, and force between the plates correspond to temperature, volume, and pressure, respectively. There exists a relationship between force, voltage, and separation which is equivalent to an equation of state. a) Describe the four steps which are analogous to the four steps in the Carnot cycle of a heat engine. Specify the direction of the cycle which is to deliver energy for mechanical purposes. b) The engine operates between two batteries which have voltages of 100 V and 20 V, respectively. With the first battery hooked up to the capacitor, the separation of the plates is changed from 0.2 m to 0.1 m. Calculate the remaining two separations corresponding to the other two corners of the cycle. c) Calculate the amount of charge absorbed, and the amount of charge emitted, for the four steps.

SOLUTION:

a) The four steps undergone by the machine are the following:

1. "isothermal change": decrease of the separation of the plates at high constant voltage (*a-b* in the figure);
2. "adiabatic change": decrease of volume while the capacitor is electrically insulated;
3. "isothermal change": increase of the separation of the plates at low constant voltage;
4. "adiabatic change": increase of volume while the capacitor is electrically insulated.

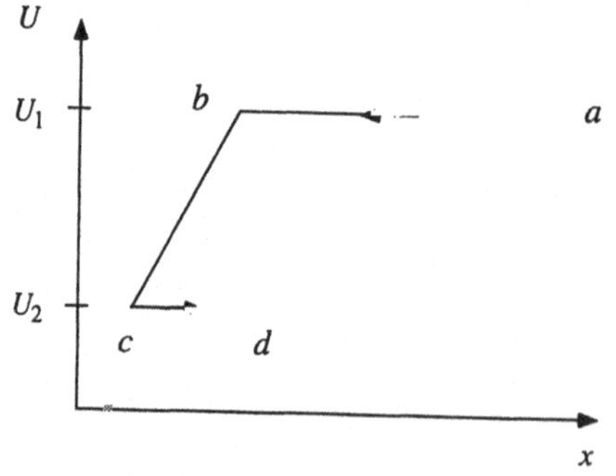

[1] See Fuchs (1986).

The "adiabats" have been calculated in Problem 8. It was found that in a process in which no electrical charge is exchanged, the voltage is proportional to the separation of the capacitor plates. Therefore the "Carnot" cycle undergone by the capacitor with variable separation can be described in the voltage-separation diagram as in the figure. The cycle is traversed counter clockwise.

b) Since voltage and separation are proportional if the separation is changed while the batteries are unhooked (steps 2 and 4), the remaining values for the separation x at points c and d are easily calculated to be $x_c = 0.02$ m and $x_d = 0.04$ m, respectively. Therefore the cycle is completely determined.

c) For steps 2 and 4 the amount of charge exchanged is zero. During the first step, charge is absorbed from the first battery at a voltage of 100 V. Since the latent charge is equal to $-\varepsilon_o A/x^2$ (see Example 5.5), the charge absorbed on a-b is determined by

$$I_q = -\Lambda_x \dot{x} - K_x \dot{U} = -\Lambda_x \dot{x}$$

$$q_e = -\int_{t_1}^{t_2} I_q dt = \int_{x_1}^{x_2} \Lambda_x dx = \varepsilon_o AU\left[\frac{1}{x_2} - \frac{1}{x_1}\right]$$

$$q_{e1} = 8.85 \cdot 10^{-12} \cdot 1.0 \cdot 100\left[\frac{1}{0.1} - \frac{1}{0.2}\right]\mathrm{C} = 4.43 \cdot 10^{-9}\,\mathrm{C}$$

The charge emitted to the battery at 20 V during the third step is calculated in the same manner; we obtain

$$q_{e3} = 8.85 \cdot 10^{-12} \cdot 1.0 \cdot 20\left[\frac{1}{0.04} - \frac{1}{0.02}\right]\mathrm{C} = -4.43 \cdot 10^{-9}\,\mathrm{C}$$

It is equal to the charge absorbed.

■

A gas which satisfies the equation of state of the ideal gas undergoes two different processes which lead from the same initial state to the same final state. Process 1 takes the gas at constant pressure from volume V_1 to volume $V_2 > V_1$. Process 2 consists of two steps: isothermal expansion at T_1 from V_1 to V_2, and isochoric heating at V_2 from temperature T_1 to T_2 (Figure 9). Calculate the energy exchanged in the mechanical processes.

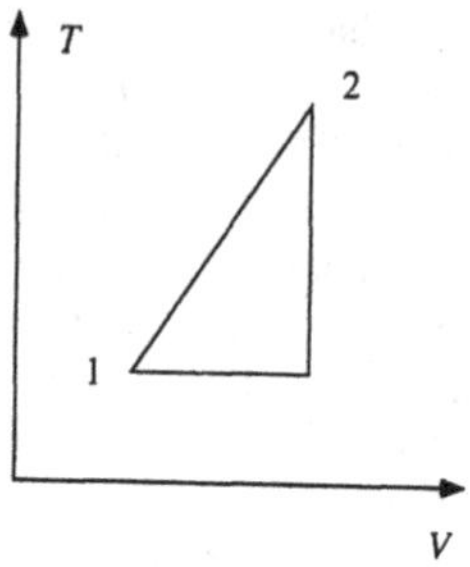

FIGURE 9. Problem 19.

SOLUTION:

1. Since the pressure stays constant, the work is calculated easily to be equal to

$$W_1 = -P \cdot (V_2 - V_1)$$

2. The volume stays constant during the second step which means that energy is exchanged in the mechanical process during the first part only. During the first step the temperature of the gas is constant and equal to T_1. Therefore:

$$W_2 = -\int P dV = -nRT_1 \ln(V_2/V_1)$$

In general, this expression is not equal to W_1!

■

If the heat exchanged in a process is to be independent of the path, the constitutive relations of the body undergoing the process *may not be arbitrary.* Calculate the heat exchanged for the processes of Problem 19 using the following two sets of constitutive relations: (a) $\Lambda_V = a / V$, $K_V = b / T$, where a and b are both constant; (b) $\Lambda_V = aT / V$, $K_V = b$.

SOLUTION:

We will perform the calculations for the ideal gas.

1. According to Equation (27) we have

$$S_e = \int [\Lambda_V dV + K_V dT]$$

(a) $\Lambda_V = a / V$, $K_V = b / T$,

$$S_e = a \ln\left(\frac{V_2}{V_1}\right) + b \ln\left(\frac{T_2}{T_1}\right)$$

(b) $\Lambda_V = aT / V$, $K_V = b$

$$S_e = \int \left[\frac{aP}{nR} dV + b dT\right] = \frac{aP}{nR}(V_2 - V_1) + b(T_2 - T_1)$$

for the process at constant pressure.

2. The two-step process yields the following results:

(a)

$$S_e = a \ln\left(\frac{V_2}{V_1}\right) + b \ln\left(\frac{T_2}{T_1}\right)$$

(b)

$$S_e = aT_1 \ln\left(\frac{V_2}{V_1}\right) + b(T_2 - T_1)$$

Only the first set (a) of constitutive relations, i.e. the ones which have been proposed for the ideal gas, lead to expressions for the heat exchanged which are independent of the path. The latter set (b) has to do with energy, and not with heat.

■

Assume that the density $\rho(T)$, the specific heat capacity $k_p(T)$ at constant pressure, and the speed of sound c in a fluid are known from experiment as functions of temperature for a given pressure. Determine from these values the ratio of the heat capacities γ, the (isothermal) compressibility κ_T, and the latent heat with respect to volume.

SOLUTION:

The results are the following:

(1) $$\gamma = 1 + \frac{\alpha_V^2 c^2}{k_P}$$

(2) $$\kappa_T = \frac{1}{\rho}\left[\frac{\alpha_V^2}{k_P} + \frac{1}{c^2}\right]$$

(3) $$\Lambda_V = \rho \frac{\alpha_V c^2 k_P}{k_p + \alpha_V^2 c^2}$$

They will be derived in turn. As you can see, they contain the temperature coefficient of expansion of the fluid which will be shown to follow from knowledge of the density as a function of temperature.
First we need some relations between density, pressure, and temperature:

$$\frac{\partial p^*}{\partial \rho} = \left[\frac{\partial \rho}{\partial P}\right]^{-1}$$

$$\frac{\partial p^*}{\partial T} = -\left[\frac{\partial \rho}{\partial P}\right]^{-1} \frac{\partial \rho}{\partial T}$$

$$\frac{\partial p}{\partial T} = \frac{\partial p^*}{\partial T}$$

These equations are derived in analogy to Equations (7). Now, the temperature coefficient of expansion (Equation 16) follows from these and from Equations (7) and (9):

$$\alpha_V = \frac{1}{V}\frac{\partial V}{\partial T} = -\frac{1}{V}\left[\frac{\partial p}{\partial V}\right]^{-1} \frac{\partial p}{\partial T}$$

$$= \frac{1}{\rho}\left[\frac{\partial p^*}{\partial \rho}\right]^{-1} \frac{\partial p^*}{\partial T} = -\frac{1}{\rho}\frac{\partial \rho}{\partial T}$$

1. Now we derive a relations between the speed of sound, the ratio of the heat capacities, and the isothermal compressibility. Equations (7) and (9) yield

$$\kappa_T = -\frac{1}{V}\frac{\partial V}{\partial P} = \frac{1}{\rho}\left[\frac{\partial p^*}{\partial \rho}\right]^{-1}$$

From Equation (70) we conclude that

$$(4) \qquad c^2 = \gamma \frac{1}{\rho \kappa_T}$$

2. Next, we shall express the difference of the heat capacities at constant pressure and at constant volume in different form. From Equations (46), (122), (17), and (19) we derive

$$K_P - K_V = -\Lambda_V \left[\frac{\partial p}{\partial V}\right]^{-1} \frac{\partial p}{\partial T}$$

$$= -\frac{\partial p}{\partial T}\left[\frac{\partial p}{\partial V}\right]^{-1} \frac{\partial p}{\partial T} = P\beta V \alpha_V$$

or

$$(5) \qquad K_P - K_V = V\frac{\alpha_V^2}{\kappa_T}$$

3. Now we eliminate the compressibility from (4) and (5) and obtain the first of the desired results (1). Remember that $k_P = K_P/m$.

4. The second result (2) is derived directly from (1) and (4).

5. The latent heat with respect to volume is equal to the derivative of pressure with respect to temperature (Equation 122), which is equal to Pb, which, from Equation (19) is equal to the ratio of the temperature coefficient of expansion and the compressibility. The compressibility was just calculated in 4.

■

Derive the energy of the ideal gas as a function of S and V. Show that you can obtain both the entropy as a function of temperature and volume, and the equation of state of the ideal gas from this information.

SOLUTION:

We know the expressions for the energy and the entropy of the ideal gas:

$$E(T) = E_o + C_V(T - T_o)$$

$$S(V,T) = S_o + nR\ln\left(\frac{V}{V_o}\right) + C_V\ln\left(\frac{T}{T_o}\right)$$

Combining these expressions, we obtain

$$E(S,V) = E_o + C_V T_o\left\{\exp\left(\frac{S - S_o}{C_V}\right)\exp\left(-\frac{nR}{C_V}\ln\left(\frac{V}{V_o}\right)\right) - 1\right\}$$

Now, according to Equation (135), the partial derivative of this function with respect to entropy and volume yield the temperature and the negative pressure:

$$(1) \qquad T = \frac{\partial E(S,V)}{\partial S} = T_o \exp\left(\frac{S - S_o}{C_V}\right)\exp\left(-\frac{nR}{C_V}\ln\left(\frac{V}{V_o}\right)\right)$$

This equation can be solved for the entropy which yields the first result:

$$S - S_o = nR\ln\left(\frac{V}{V_o}\right) + C_V\ln\left(\frac{T}{T_o}\right)$$

If we also calculate the second derivative

$$(2) \qquad P = -\frac{\partial E(S,V)}{\partial V} = \frac{nR}{V} T_o \exp\left(\frac{S - S_o}{C_V}\right)\exp\left(-\frac{nR}{C_V}\ln\left(\frac{V}{V_o}\right)\right)$$

Combining (1) and (2) demonstrates that

$$PV = nRT$$

■

In Section I.1.2, the compressibility and the thermal coefficient of expansion were defined. Prove the following relationship:

$$K_p - K_V = V\frac{\alpha^2}{\kappa_T}$$

Transform this relation to show that the specific temperature coefficients of enthalpy and of energy are related by

$$c_p = c_V\left(1+\gamma^*\alpha_V T\right)$$

where

$$\gamma^* = \frac{\alpha_V}{\rho c_V \kappa_T} = \frac{\alpha_V}{\rho c_p \kappa_s}$$

is called the Grüneisen ratio. (α_V and κ_T are the temperature coefficient of volume and the isothermal compressibility, respectively.) Show that for the ideal gas the Grüneisen ratio is

$$\gamma^* = \frac{c_p}{c_V} - 1 = \gamma - 1$$

Hint: Use the relation between the different coefficients defined according to Section I.1.2 and remember the Carnot-Clapeyron law derived in Section I.3.4.

SOLUTION:

1. The first of the relations was derived in Problem 21 already.

2. From the first relation we obtain

$$TK_p - TK_V = TV\frac{\alpha_V^2}{\kappa_T} = \frac{m}{\rho}T\frac{\alpha_V^2}{\kappa_T}$$

$$\Rightarrow \qquad c_P - c_V = T\frac{\alpha_V^2}{\rho\kappa_T}$$

$$\Rightarrow \qquad c_P = c_V + T\frac{\alpha_V^2}{\rho\kappa_T} = c_V\left(1+T\frac{\alpha_V^2}{c_V\rho\kappa_T}\right)$$

$$= c_V\left(1+\gamma^*\alpha_V T\right)$$

■

Why should the prejudice that the "heat" of an ideal gas is the energy of the motion of its particles lead to the prediction of constant "heat capacity" of a monatomic gas? Why should we reject the caloric theory of heat if the "heat capacity" of the ideal gas were indeed constant? Why does the notion of the motion of the least particles being the "heat" of the gas contradict the very theory (the mechanical theory of heat) which was proposed in place of the caloric theory?

SOLUTION:

1. In Problem 47 of Chapter 2, we derived an expression for the molar temperature coefficient of energy ("heat capacity") for the monatomic ideal gas:

$$\bar{c}_V = \frac{3}{2}R$$

having assumed that the energy of the gas is the kinetic energy of the particles. This quantity is constant.

2. Equation (79) demonstrates that the heat capacities of the caloric theory cannot be constant.

3. In the mechanical theory of heat, "heat" is the energy *transferred* in heating or cooling. "Heat" as an energy form does *not* reside in the gas.

■

CHAPTER 3

Solutions of Selected Problems

An immersion heater in a water kettle is hooked up to 220 V. Its electrical resistance is 160 Ω at a temperature of 20°C; the temperature coefficient of the resistance is $4 \cdot 10^{-3}\ K^{-1}$. If the heat transfer coefficient between heater and water is 100 W/(K · m²) and the surface area of the heater is 0.020 m², how large will the energy current from the heater to the water be? How does the situation change if a layer of mineral deposit builds up around the heater?

SOLUTION:

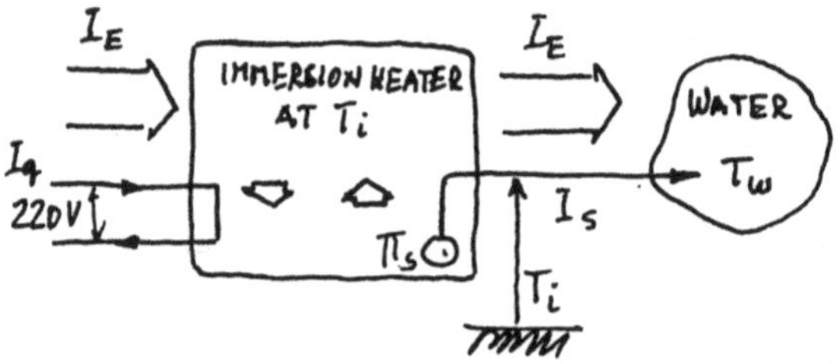

Resistance law: $I_q = U/R$

Resistance: $R = R_0(1+\alpha(T-T_0))$

El. energy current: $I_E = U^2/R$

Thermal energy current: $I_E = hA(T_i - T_w)$

The energy currents are equal: $$hA(T_i - T_w) = \frac{U^2}{R_0(1+\alpha(T_i - T_0))}$$

Solution of the quadratic equation for $T_w = 40°C$:

	$h = 100\ W/(K\cdot m^2)$	$h = 70\ W/(K\cdot m^2)$
Temperature of immersion heater	142°C	174°C
Energy current	203 W	187 W

The buildup of a mineral deposit leads to a reduction of the overall heat-transfer coefficient h. The immersion heater responds by increasing its temperature, increasing the resistance which decreases I_E. ■

Show that the energy current transmitted through a cylindrical shell of length L having inner and outer radii r_1 and r_2 is

$$I_E = \pi L\left[\frac{1}{2r_1h_1}+\frac{1}{2r_2h_2}+\frac{1}{2k_E}\ln\left(\frac{r_2}{r_1}\right)\right]^{-1}(T_1-T_2)$$

where h_1 and h_2 are the inner and the outer convective heat transfer coefficients. The temperatures of the fluids on the inside and the outside are T_1 and T_2.

SOLUTION:

$$|I_E| = |\Delta T|/R_E$$

$$R_E = R_{E,1} + R_{E,shell} + R_{E,2}$$

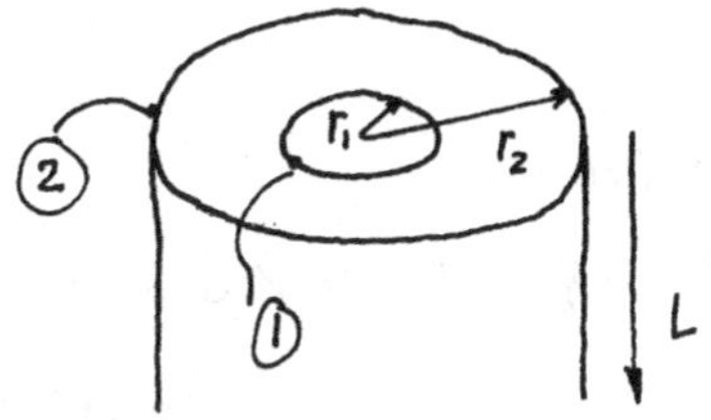

Convective heat transfer at boundaries:

$$R_{E,1} = 1/(h_1 A_1) = 1/(h_1 2\pi r_1 L)$$

$$R_{E,2} = 1/(h_2 A_2) = 1/(h_2 2\pi r_2 L)$$

Thermal resistance of cylindrical shell (see Example 5 for calculation in the case of a spherical shell):

$$R_{E,shell} = \int_{r_1}^{r_2}\frac{dr}{k_E A(r)} = \int_{r_1}^{r_2}\frac{dr}{2\pi r L k_E} = \frac{1}{2\pi L k_E}\ln(r_2/r_1)$$

$$\Rightarrow \quad I_E = R_E^{-1}(T_2-T_1) = 2\pi L\left[\frac{1}{r_1h_1}+\frac{1}{r_2h_2}+\frac{1}{k_E}\ln\left(\frac{r_2}{r_1}\right)\right]^{-1}(T_1-T_2)$$

■

A cylindrical volume of rock below ground has been heated uniformly to 50°C while the rest of the rock has a temperature of 10°C. (This might be done in solar seasonal heat storage applications.) For properties of the rock use the average values for granite from the tables in the Appendix. a) For heat loss from the cylindrical area to the surroundings make the following model. While the temperatures of the storage area and the surroundings remain uniform, heat flows through a cylindrical mantle with inner and outer radii equal to half and to twice the radius of the storage cylinder, respectively. Estimate the energy current due to heat loss for a radius of 5.0 m and a length of the cylindrical space of 40 m. b) How large should the radius be made for heat loss over a period of half a year not to exceed one quarter of the energy stored in the cylinder?

SOLUTION:

Table A.9: $k_E \approx 2.5\ W/(K \cdot m)$

$\rho = 2640\ kg/m^3$

$c_p = 820\ J/(K \cdot kg)$

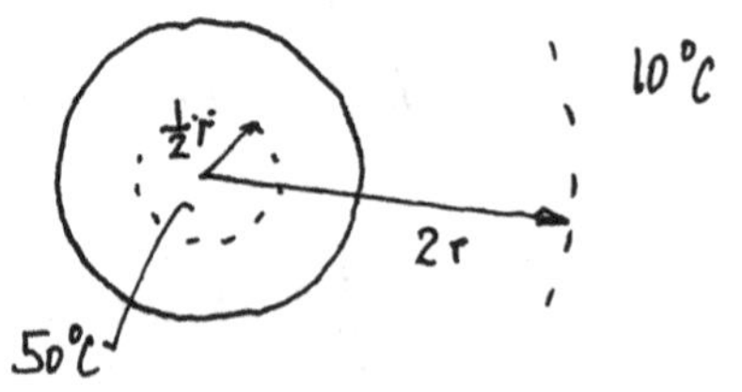

a) See Problem 4: $I_E \approx \Delta T / R_E = 2\pi L k_E \left[\ln(r_2/r_1)\right]^{-1} (T_1 - T_2)$

$= 2\pi \cdot 40 \cdot 2.5 \left[\ln(4)\right]^{-1} (50 - 10)\ W = 18.1\ kW$

This value holds at the beginning, with $T_1 = 50°C$.

b) RC-model: $E(t) = E_0 e^{-t/\tau}$ with the energy $E = 0$ at $T_2 = 10°C$.

$$\tau = R_E \cdot C = R_E \cdot \pi r^2 L \rho c_p \quad \text{with } R_E = 2.2 \cdot 10^{-3}\ K/W$$

$$E(t = 1/2\ a) = 3/4\ E_0 \Rightarrow 3/4\ E_0 = E_0 e^{-t_{1/2}/\tau} \rightarrow \tau = t_{1/2} / \ln(4/3)$$

$$\rightarrow r = \sqrt{\frac{t_{1/2}}{\ln(4/3) R_E \pi L \rho c_p}} = \sqrt{\frac{16 \cdot 10^6\ m^2}{\ln(4/3) \cdot 2.2 \cdot 10^{-3} \pi\, 40 \cdot 2640 \cdot 820}} = 9.6\ m$$

■

With the definition of the absolute brightness of stars given in magnitudes (Problem 7), show that a star's radius R and its surface temperature T are related by

$$M = 42.3 - 5\log_{10}(R/R_s) - 10\log_{10}(T)$$

if we take it to radiate as a black body.

SOLUTION:

We need the last result of Problem 7 and values for the Sun (p. 416-417):

$$M = M_s - 2.5\log_{10}(L/L_s)$$

$$M_s = 4.79\ , \quad L_s = 3.844\cdot 10^{26}\,W\ , \quad R_s = 6.96\cdot 10^{8}\,m$$

Black body radiation from surface of spherical Sun and star:

$$L = 4\pi R^2 \sigma T^4 \quad \rightarrow \quad T_s = \left(\frac{L_s}{4\pi R_s^2 \sigma}\right)^{1/4} = 5777\,K$$

$$M = 4.79 - 2.5\log_{10}\left(\frac{4\pi\sigma R^2 T^4}{4\pi\sigma R_s^2 T_s^4}\right)$$

$$= 4.79 - 2.5\log_{10}\left(\frac{1}{T_s^4}\left(\frac{R}{R_s}\right)^2 T^4\right)$$

$$= 42.4 - 2.5\left\{\log_{10}(R/R_s)^2 + \log_{10}(T^4)\right\}$$

$$= 42.4 - 5\log_{10}(R/R_s) - 10\log_{10}(T)$$

It is customary to express stellar radii in terms of the radius of the sun, whereas temperatures are given in absolute values. ■

A spherical satellite orbits the Earth at a distance of twice the Earth's radius from the center of the planet. Estimate its temperature in the shadow of our planet if you assume the surface of the satellite not to be selective.

SOLUTION:

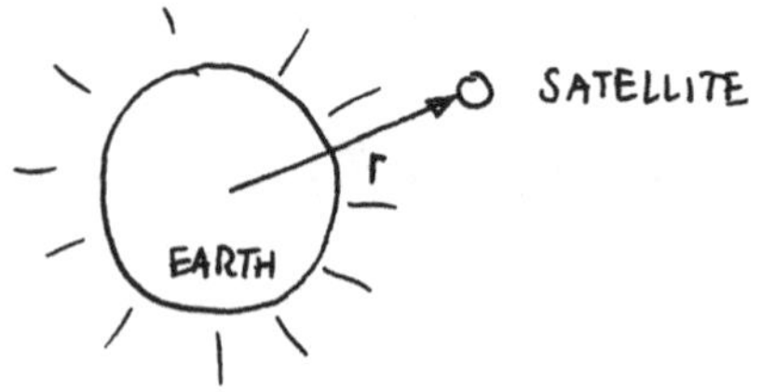

Steady-state balance of energy for satellite:

$$\pi R_s^2 \cdot G_E = 4\pi R_s^2 \sigma T_s^4$$

G_E: irradiance due to radiation from Earth:

$$G_E = \frac{1}{4\pi r^2} 4\pi R_E^2 \sigma T_E^4 \qquad T_E = 279\ K \text{ (Example 10)}$$

$$= \frac{\sigma (279)^4}{4}\ W/m^2 = 86\ W/m^2$$

$$\rightarrow \quad T_s = \left(\frac{G_E}{4\sigma}\right)^{1/4} = 140\ K$$

■

A sheet of metal with a selective surface of 2.0 m² is lying horizontally on the ground. The bottom side of the sheet is well insulated. In the visible part of the spectrum the emission coefficient of the metal is 0.90, while in the infrared it is 0.30. Take the ambient temperature to be 20°C. The Sun stands 50° above the horizon, and 70% of the radiation outside the atmosphere penetrates the air. (Assume all the radiation from the sky to be direct and not diffuse.) a) Neglecting convection, how large should the temperature of the metal sheet be in the light of the Sun? b) Now take into consideration convective heat transfer at the surface of the sheet. The convective heat transfer coefficient is assumed to be 14 W/(K · m²). Calculate the temperature attained by the sheet under these conditions.

SOLUTION:

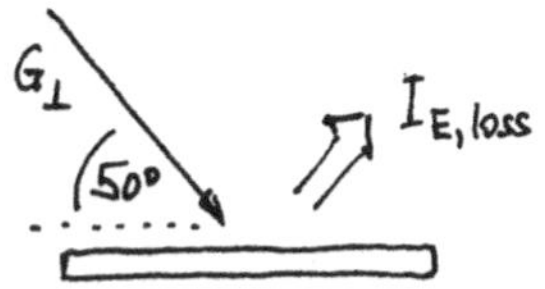

$$G_\perp = 0.70 \cdot G_{sc} = 952 \text{ W/m}^2$$
$$G = G_\perp \sin(50°) = 729 \text{ W/m}^2$$

a) Steady-state balance of energy:

$$a_v A G = I_{E,loss}$$
$$I_{E,loss} = A e_i \sigma (T_s^4 - T_a^4)$$

$$\Rightarrow 0.90 \cdot 729 \text{ W/m}^2 = 0.30\,\sigma\left(T_s^4 - (293\text{ K})^4\right) \Rightarrow T_s = 463\text{ K}$$

b) Steady-state balance of energy including convection:

$$a_v A G = A e_i \sigma (T_s^4 - T_a^4) + A h (T_s - T_a)$$

Numerical solution of the nonlinear equation yields $T_s = 334$ K, which is much lower and considerably more realistic.

■

In solar energy applications, parabolic troughs are used to focus light upon absorbers of cylindrical shape. Calculate the heat loss coefficient of such an absorber. Consider it to be made of a metal pipe of diameter of 5.5 cm, surrounded by a thin glass cover with an outer diameter of 8.5 cm. The annulus between the pipe and the cover is evacuated. Take the convective heat transfer coefficient at the surface of the cover to be 35 W/(K · m²). The emissivities of glass and the metal pipe are 0.88 and 0.92, respectively. Present the result as a function of absorber temperature for an ambient temperature of 20°C.

SOLUTION:

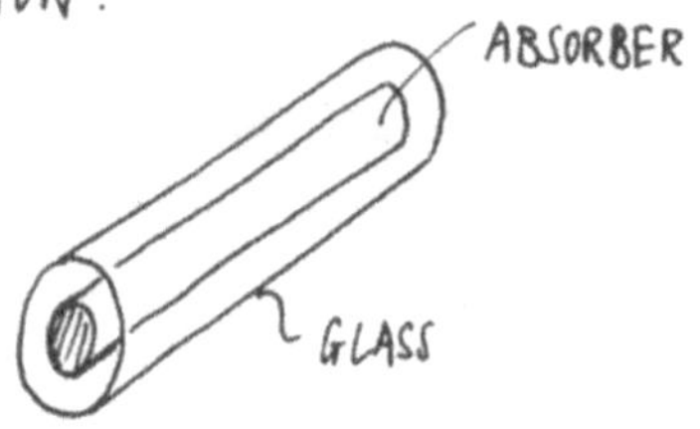

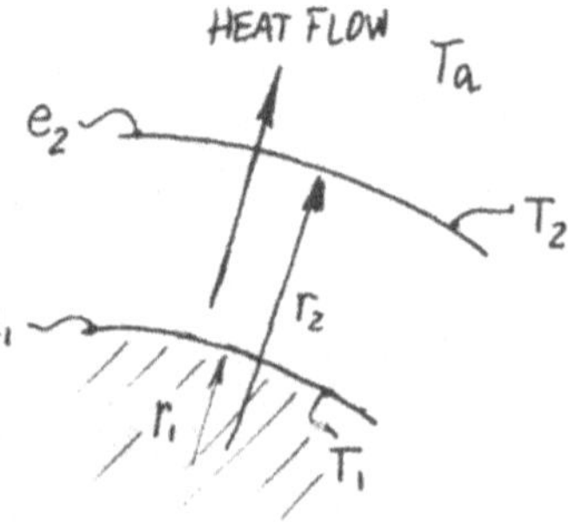

There is transport of entropy and energy from the absorber to the glass due to radiation, and from the glass tube to the surroundings as a result of convection and radiation. The energy current from the absorber to the cover is given by an expression similar to Equation (46):

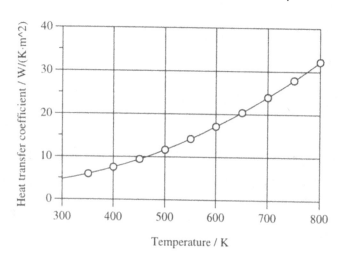

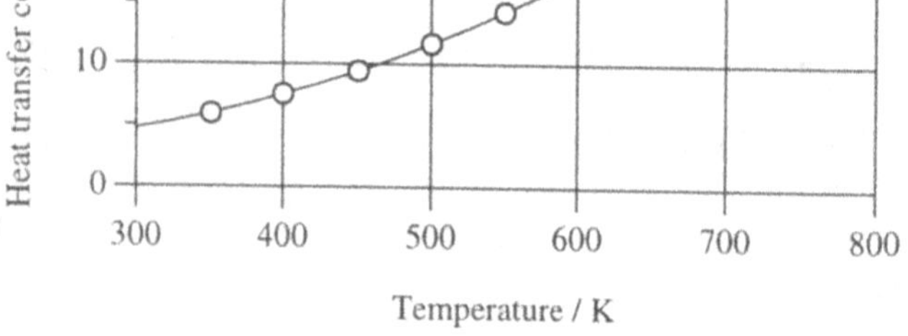

$$I_E = \frac{A_1 \sigma (T_1^4 - T_2^4)}{1/e_1 + (1/e_2 - 1) A_1/A_2}$$

The same energ current also flows to the environment:

$$I_E = A_2 \left[h(T_2 - T_a) + \sigma e_2 (T_2^4 - T_a^4) \right]$$

These equations have to be solved for T_2 and I_E, with T_1, T_a, and the material parameters specified. We can finally define the overall heat transfer coefficient h_{tot} with respect to the absorber:

$$I_E = A_1 h_{tot} (T_1 - T_a)$$

The results are presented in the graph on the previous page. The heat transfer coefficient increases strongly with temperature. While the losses are large, so is the energy flux of incident radiation at a concentration ratio of about 40 (see Section 3.6.3). In steady state, we can get temperatures of 700 K to 800 K.

Radiative energy transfer between absorber and cover (see also page 331):

I_{E1} : emitted + reflected part of part of I_{E2}

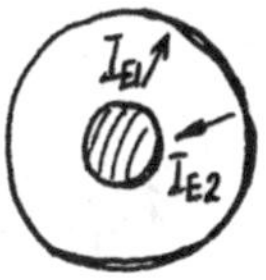

1) $I_{E1} = A_1 e_1 \sigma T_1^4 + (1 - e_1) A_1/A_2 \, I_{E2}$

A_1 intercepts a fraction A_1/A_2 of I_{E2}

2) $I_{E2} = A_2 e_2 \sigma T_2^4 + (1 - e_2) \{ I_{E1} + (1 - A_1/A_2) I_{E2} \}$

A_2 intercepts all of I_{E1} plus the fraction $1 - A_1/A_2$ from other parts of A_2.

Net energy current leaving A_1 : $I_E = I_{E1} - \frac{A_1}{A_2} I_{E2}$. This delivers the equation used above.

■

A spherical thin-walled water tank has a volume of 1.0 m³. The water inside is kept at a constant temperature of 60°C by heating it with an energy current equal to 1.0 kW. The ambient temperature is 15°C. How long will it take for the water to reach a temperature of 40°C after the heater has been turned off?

SOLUTION:

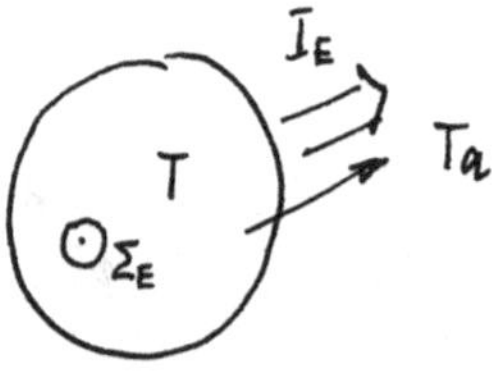

The heat transfer coefficient can be determined from steady-state conditions:

$$I_E = \Sigma_E$$

$$I_E = hA\,(T_w - T_a)$$

$$h = \frac{\Sigma_E}{A\,(T_w - T_a)} = \frac{\Sigma_E}{4\pi r^2 (T_w - T_a)} = \frac{1000}{4\cdot\pi\ 0.62^2\cdot 45}\ \frac{W}{K\cdot m^2}$$

$$= 4.60\ W/K\cdot m^2$$

RC-model of cooling: $T(t) = T_a + (T_i - T_a)\,e^{-t/\tau}$

$$\tau = R_E \cdot C = 1/(hA)\cdot \rho V c_p$$

$$= 0.045\cdot 1000\cdot 1.0\cdot 4200 = 1.9\cdot 10^5 s$$

$$e^{-t/\tau} = \frac{T(t) - T_a}{T_i - T_a} = \frac{40-15}{60-15} = 0.556$$

$$\Rightarrow\ t_{cooling} = 1.1\cdot 10^5 s \approx 1.3\ \text{days}$$

■

A body of water having a volume of 1.0 m³ loses heat to the surroundings. The temperatures are 80°C and 20°C for the water (initially) and the environment, respectively. The product of total heat transfer coefficient and surface area is 60 W/K. a) How long does it take for the temperature difference between the water and the surroundings to decrease to half its initial value? b) How large is the rate of production of entropy right at the beginning? c) How much entropy is produced in total from the beginning until the water has cooled down completely? d) How much energy could have been released by an ideal Carnot engine operating between the water and the environment as the water cools to ambient temperature?

SOLUTION:

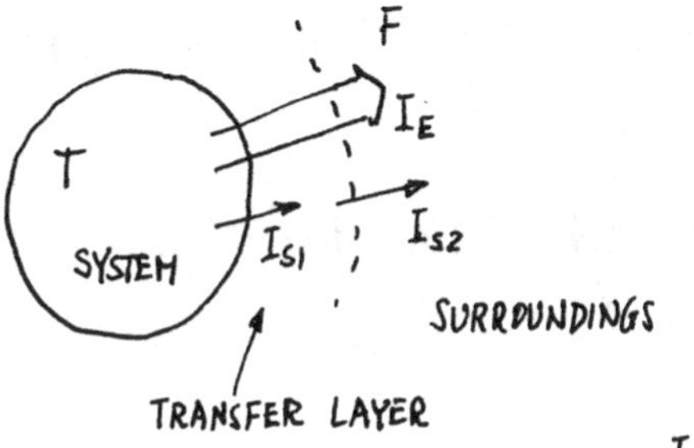

a) RC-model of cooling:

$$\left.\begin{aligned} \dot{E} &= -I_E \\ I_E &= hA(T - T_a) \\ \dot{E} &= C\dot{T} = mc\dot{T} \end{aligned}\right\} \quad mc\dot{T} = -hA(T - T_a)$$

Solution of DE:

$$\frac{mc}{hA}\int_{T_i}^{T_f} \frac{dT}{T - T_a} = -\int_0^t dt$$

$$\rightarrow \quad t = -\frac{mc}{hA}\ln\left(\frac{T_f - T_a}{T_i - T_a}\right) = -\frac{1000\cdot 4200}{60}\ln\left(\frac{50-20}{80-20}\right)\text{s} = 4.85\cdot 10^4\,\text{s}$$

b) Entropy production in the transfer layer:

$$\pi_S = |I_{S2}| - |I_{S1}| \; ; \quad |I_{S1}| = |I_E|/T \; ; \quad I_{S2} = |I_E|/T_a$$

$$\rightarrow \quad \pi_S = |I_{Ei}|\left(1/T_a - 1/T_i\right) = 60\cdot(80-20)\left(1/293 - 1/353\right)\text{W/K} = 2.09\,\text{W/K}$$

Different approach: Consider transfer layer as part of the system:

Balance of entropy: $\Pi_s = \dot{S} + I_{s2} \Rightarrow \Pi_s = C\dot{T}/T + I_{s2}$

$$\Pi_s = -I_E/T + I_{s2}$$

c) Integrated form of balance of entropy for system plus environment:

$$\Delta S = S_{prod} \quad , \quad \Delta S = \Delta S_w + \Delta S_{environment}$$

$$\Delta S_w = mc \ln\left(T_f/T_i\right) \; ; \; \Delta S_{env} = S_{e,env} = Q/T_a = mc\left(T_i - T_f\right)/T_a$$

$$\Rightarrow S_{prod} = mc \ln\left(\frac{T_f}{T_i}\right) + \frac{1}{T_a} mc\left(T_i - T_f\right) = -7.82 \cdot 10^5 \,J/K + 8.60 \cdot 10^5 \,J/K$$

$$= 0.78 \cdot 10^5 \,J/K$$

d)

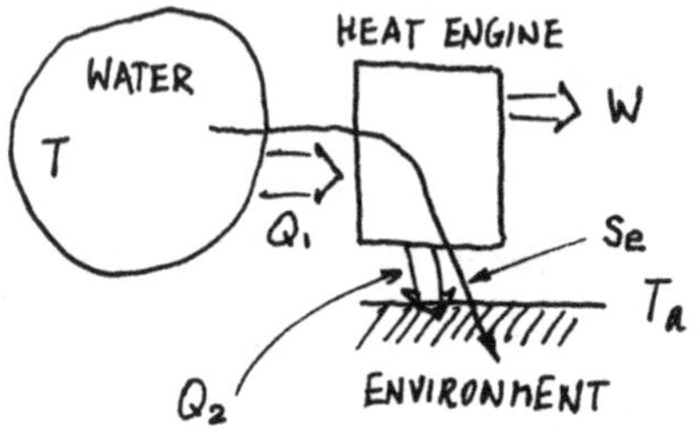

$$|W| = |Q_1| - |Q_2|$$

$$= mc\left(T_i - T_f\right) - T_a \cdot S_e$$

$$= mc\left(T_i - T_f\right) - T_a \, mc \ln\left(T_i/T_f\right)$$

$$= 1000 \cdot 4200 \left[60 - 293 \ln\left(\frac{353}{293}\right)\right]$$

$$= 22.7 \text{ MJ}$$

In the actual process, this energy is "lost". Therefore, we should also obtain the result by multiplying the entropy produced and the temperature of the environment: $|W|_{lost} = T_a \cdot S_{prod}$.

■

To maintain an inner temperature of 20°C in a building situated in a 0°C environment, the required heating load is 5 kW. Without heating, the house is found to cool down as follows: every day, its temperature decreases by 1/5 of the temperature difference to the environment. a) Determine the product of surface area and total heat transfer coefficient. b) Model the building as a single node system. Calculate its temperature coefficient of energy. c) Assume the temperature inside the building to be 12°C. Calculate the heating power necessary if you wish the temperature to rise by 1°C per hour.

SOLUTION :

a)

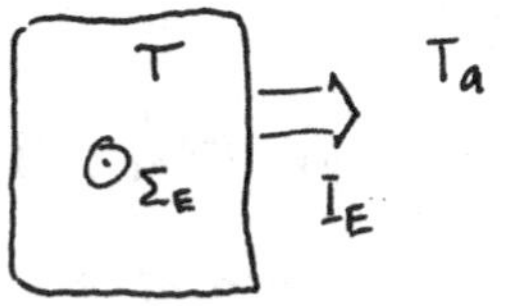

Steady-state balance of energy for the building:

$$\dot{E} + I_E = \Sigma_E \quad , \quad \dot{E} = 0$$

$$|I_E| = hA\,(T - T_a)$$

$$\Rightarrow \quad hA = \frac{\Sigma_E}{T - T_a} = \frac{5000\,\text{W}}{(20-0)\,\text{K}} = 250\ \text{W/K}$$

b) RC-model of cooling. Solution: $T(t) - T_a = (T_i - T_a)\,e^{-t/\tau}$

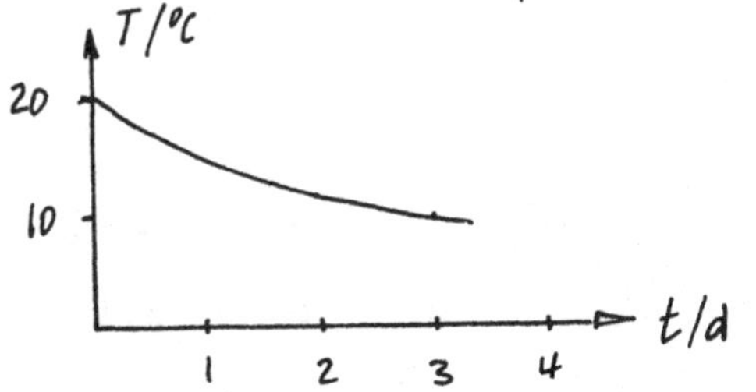

where $\tau = R_E \cdot C$, $R_E = 1/hA$

$$(16-0)\,\text{K} = (20-0)\,\text{K}\,e^{-86400\,\text{s}/\tau}$$

$$\rightarrow \quad \tau = 3.87 \cdot 10^5\,\text{s} \approx 4.5\,\text{d}$$

$$R_E \cdot C = 1/hA \cdot C = 3.87 \cdot 10^5\,\text{s} \quad \rightarrow \quad C = 9.7 \cdot 10^7\ \text{J/K}$$

c)

$$\left.\begin{aligned} \dot{E} + I_E &= \Sigma_E \\ \dot{E} &= C\dot{T} \end{aligned}\right\} \quad \Sigma_E = C\dot{T} + I_E = C\dot{T} + hA\,(T - T_a)$$

$$= 9.7 \cdot 10^7 \frac{1}{3600}\,\text{W} + 250\,(12-0)\,\text{W} = 30\,\text{kW}$$

■

An initially empty tank is filled with an externally heated fluid. Consider the irreversible processes due to fluid flow and heat loss from the tank to the environment. Take fluid friction to obey the law of Hagen and Poiseuille. Heat loss from the tank should be proportional to how much hot fluid is in the tank. a) Give qualitative reasons to show that there should be an optimal rate of charging of the tank. b) Show that under these conditions, the optimal charging time should be proportional to the square root of the frictional resistance (as in Ohm's law, see the Prologue), and inversely proportional to the square root of both the total heat transfer coefficient and the difference of temperatures between the hot fluid and the environment. (Assume the temperature decrease of the fluid in the tank due to cooling to be small; i.e., take the temperature of the heated fluid to remain constant.)

SOLUTION:

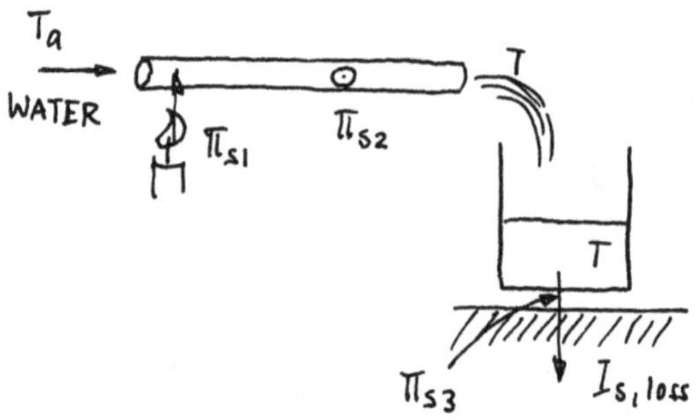

Π_{S1} : entropy production due to burning of fuel

Π_{S2} : friction

Π_{S3} : heat transfer

Tank has to be charged to volume V.

a) If we make the charging time Δt smaller, we increase the volume flux and therefore friction in the pipe (Π_{S2} increases). On the other hand, the water is in the tank for a shorter duration, leading to a decrease of heat loss and therefore to less entropy being produced. (The third source of entropy production, burning of fuel, does not appear to be influenced by this consideration: we need to produce a fixed amount of entropy to increase the temperature of the water from T_a to T.).

b) Since we are interested in the overall process lasting a time span Δt, we will consider the integrated form of the entropy production:

$$S_{prod} = S_{prod,1} + S_{prod,2} + S_{prod,3}$$

1. $S_{prod,1} = \Delta S_{Water}$ due to heating of volume V from T_a to T
$= c \rho V \ln(T/T_a)$

2. $S_{prod,2} = \Pi_{S2} \Delta t = \frac{1}{T} \mathcal{Q} \Delta t = \frac{1}{T} R_v I_v^2 \Delta t = \frac{1}{T} R_v V^2 \frac{1}{\Delta t}$

Here we neglect the possible rise of T due to friction.

3. $S_{prod,3} = \overline{\Pi}_{S3} \Delta t \; ; \quad \overline{\Pi}_{S3} = h \bar{A} (T - T_a)(1/T_a - 1/T)$

$\bar{A}$ is the average surface area loosing heat

$\bar{A} = \frac{1}{2} b V$

$$\Rightarrow S_{prod} = c \rho V \ln(T/T_a) + \frac{1}{T} R_v V^2 \frac{1}{\Delta t} + \frac{1}{2} b V h (T - T_a)(1/T_a - 1/T) \Delta t$$

$$\frac{\partial S_{prod}}{\partial \Delta t} = -\frac{1}{T} R_v V^2 \frac{1}{\Delta t^2} + \frac{1}{2} b V h \frac{(T - T_a)^2}{T T_a}$$

$$\frac{\partial S_{prod}}{\partial \Delta t} = 0 \Rightarrow \Delta t^2_{optimum} = \frac{1}{T} R_v V^2 \frac{2}{b V h} \frac{T T_a}{(T - T_a)^2}$$

$$\Delta t_{optimum} \sim \sqrt{R_v} \frac{1}{\sqrt{h}} \frac{1}{T - T_a}$$

■

According to Table A.9, the conductivity with respect to entropy of water depends less upon temperature than its counterpart, the conductivity with respect to energy. Taking the former quantity as constant for steady-state conduction through a slab of water a) should the temperature gradient be steeper at the hotter or at the cooler side? b) Show that the field equation for temperature should take the form

$$T\frac{d^2T}{dx^2}+\left(\frac{dT}{dx}\right)^2=0$$

SOLUTION:

a) Fourier's law: $|j_s| = k_s \left|\frac{dT}{dx}\right|$ where k_s is taken to be constant.

Since entropy is produced in conduction, the entropy current must increase from the hotter to the cooler end:

$$|j_{s2}| > |j_{s1}| \rightarrow |dT/dx|_2 > |dT/dx|_1$$

b) Balance of entropy: $\frac{dj_s}{dx} = \pi_s$ (Equation 101)

Fourier's law: $j_s = -k_s\frac{dT}{dx}$ (Equation 13)

Rate of entropy production: $\pi_s = -\frac{1}{T}j_s\frac{dT}{dx}$ (Equation 106)

$$\rightarrow \frac{d}{dx}\left(-k_s\frac{dT}{dx}\right) = -\frac{1}{T}\left(-k_s\frac{dT}{dx}\right)\frac{dT}{dx}$$

$$\rightarrow \frac{d^2T}{dx^2}+\frac{1}{T}\left(\frac{dT}{dx}\right)^2 = 0 \quad \text{since } k_s = \text{const.}$$

■

Consider the conduction of heat through the Earth's crust, whose geometry can be taken as flat. Allow for sources of entropy in the material which are assumed to be distributed evenly, and let the conductivity with respect to energy be constant. a) Show that the temperature profile from the base of the crust to the surface is

$$T(x) = T_L + \frac{1}{2}\frac{\sigma_E}{k_E}\left(L^2 - x^2\right) + \frac{1}{k_E} j_E(0)(L - x)$$

for a given energy flux $j_E(0)$ at the base and surface temperature T_L. (The thickness of the crust is L.) b) Determine the dependence of the temperature gradient near the surface upon the conductivity, the energy flow at the base, and the source rate of energy in the material. c) Calculate the surface temperature gradient for a thickness of the crust of 50 km, a thermal conductivity of 2.5 W/(K · m), and a source rate of $1.25 \cdot 10^{-6}$ W/m³.

SOLUTION:

Equation (119): $\frac{d}{dx}\left(k_E \frac{dT}{dx}\right) = -\sigma_E$

a) $k_E = \text{const} \rightarrow d^2T/dx^2 = -\sigma_E/k_E$

First integration: $\frac{dT}{dx} = -\frac{\sigma_E}{k_E} x + C_1$

BC at $x = 0$: $-k_E \frac{dT}{dx}\Big|_{x=0} = j_E(0)$

$\Rightarrow \frac{dT}{dx} = -\frac{\sigma_E}{k_E} x - \frac{1}{k_E} j_E(0)$

Second integration: $T(x) = -\frac{1}{2}\frac{\sigma_E}{k_E} x^2 - \frac{1}{k_E} j_E(0)\, x + C_2$

BC at $x = L$: $T(L) = T_L \Rightarrow C_2 = T_L + \frac{1}{2}\frac{\sigma_E}{k_E} L^2 + \frac{1}{k_E} j_E(0) \cdot L$

b) $dT/dx|_L = -\sigma_E/k_E \cdot L - 1/k_E\, j_E(0)$

c) With $\sigma_E = 1.25 \cdot 10^{-6}$ W/m³, $j_E(0) = 0$ delivers the energy flux calculated in Example 5. $\rightarrow dT/dx|_L = -(1.25 \cdot 10^{-6}/2.5) \cdot 50 \cdot 10^3$ K/m $= -0.025$ K/m. ∎

Consider the transport of heat with radiation in the interior of a star which we model as being in spherically symmetric hydrostatic equilibrium; changes of volume of the gas are assumed not to disturb this situation. Nuclear reactions release energy, with the source rate given by the specific rate $\sigma_{E,r}$ (i.e., the rate divided by the mass). The luminosity $L(r)$ is the total energy flux penetrating the spherical surface at radius r. a) Model stellar matter as a simple fluid and show that the rate of change of the specific entropy s (entropy per mass) must be given by

$$T\dot{s} = \sigma_{E,r} - \frac{\partial L}{\partial m}$$

Here, the independent variable has been changed to the mass $m(r)$ inside the sphere of radius r. b) Show that the gradient of the luminosity $\partial L / \partial m$ is given by

$$\frac{\partial L}{\partial m} = \sigma_{E,r} - \frac{3}{2}\rho^{2/3}\frac{d}{dt}\left(\frac{P}{\rho^{5/3}}\right)$$

for a monatomic ideal gas. c) Show that the gradient of luminosity inside a star is determined by the source rate due to reactions only if steady-state conditions prevail.

SOLUTION:

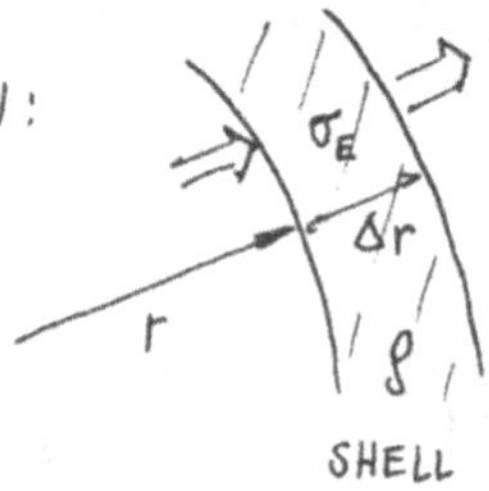

a) Consider the balance of energy of the fluid in the shell:

$$\dot{E} = \Sigma_{E,r} - (I_{E,in} + I_{E,out}) - P\dot{V}$$

While changes of density are allowed, resulting gravitational processes are disregarded. $\Sigma_{E,r}$: energy source rate due to reactions.

Gibbs fundamental form: $\dot{E} = T\dot{S} - P\dot{V}$

$$\Rightarrow \quad T\dot{S} = \Sigma_{E,r} - (I_{E,in} + I_{E,out})$$

$$\Rightarrow \quad T\dot{s} = \sigma_{E,r} - \frac{I_{E,in} + I_{E,out}}{4\pi r^2 \Delta r \cdot \rho} \quad \Rightarrow \quad T\dot{s} = \sigma_{E,r} - \frac{1}{4\pi r^2 \rho}\frac{\partial L}{\partial r}$$

Change to mass as independent variable:

$$\frac{\partial m}{\partial r} = 4\pi r^2 \rho \quad \Rightarrow \quad T\dot{s} = \sigma_{E,r} - \frac{\partial L}{\partial m}$$

b) From the last result we have $\partial L/\partial m = \sigma_{E,r} - T\dot{s}$. We now need the rate of change of the specific entropy expressed in terms of P and ρ. We can start, for example, with Equation (73) in the Interlude:

$$\dot{s} = \left(\frac{\partial P(V,T)}{\partial T}\right)^{-1} K_V \dot{P} - K_P \frac{\partial P(\rho,T)}{\partial \rho} \left(\frac{\partial P(V,T)}{\partial T}\right)^{-1} \dot{\rho}$$

Equ. (2.29) + (2.30): $\quad \partial P/\partial T = nR/V \;, \quad \partial P/\partial \rho = R/M_0 \cdot T$

Equ. (2.71) + (2.70): $\quad K_V = n\frac{3}{2}R/T \;, \quad K_P = n\frac{5}{2}R/T$

$$\dot{s} = \left(\frac{nR}{V}\right)^{-1} \frac{3}{2} nR \frac{1}{T} \dot{P} - \frac{5}{2} nR \frac{1}{T} \frac{R}{M_0} T \left(\frac{nR}{V}\right)^{-1} \dot{\rho} = \frac{m}{\rho} \frac{3}{2} \frac{1}{T} \dot{P} - \frac{m}{\rho} \frac{5}{2} \frac{R}{M_0} \dot{\rho}$$

$$\rightarrow \quad \dot{s} = \frac{3}{2} \frac{R}{M_0} \frac{1}{P} \dot{P} - \frac{5}{2} \frac{R}{M_0} \frac{1}{\rho} \dot{\rho} = \frac{3}{2} \frac{R}{M_0} \frac{\rho^{5/3}}{P} \frac{d}{dt}\left(\frac{P}{\rho^{5/3}}\right)$$

from which we get the desired result for $\partial L/\partial m$ as stated in the problem.

c) According to the result just derived, $\partial L/\partial m = \sigma_{E,r}$ for steady-state conditions. In this case, nuclear reactions lead to the radial increase of the energy flux L.

■

Calculate the normal spectral irradiance for solar radiation at the distance of the Earth for a blackbody spectrum of temperature 5777 K. (Normal means for a plane perpendicular to solar rays.)

SOLUTION: The irradiance would generally be computed from the intensity of the radiation with the help of Equation (142):

$$G = \int_{\text{solid angle}} i_E \cos(\theta)\, d\Omega$$

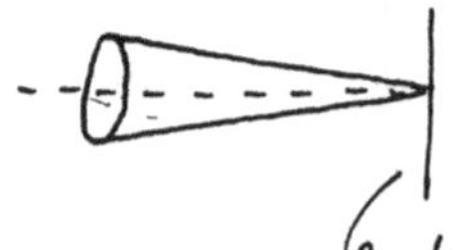

where θ is the angle between the surface normal vector and the radiation travelling toward the surface. Now, for solar radiation, the total solid angle is small, and for a surface perpendicular to the radiation, $\theta = 0$. $\rightarrow$ $G_\perp = \Omega_s \cdot i_E$

Spectral intensity: $i_{E\lambda} = \dfrac{2hc^2}{\lambda^5} \dfrac{1}{e^{hc/k\lambda T} - 1}$ (Equ. 18 in Example 35)

$$\rightarrow G_{\lambda\perp} = \Omega_s \cdot i_{E\lambda} = \pi \frac{R_s^2}{d^2} \frac{2\pi hc^2}{\lambda^5} \frac{1}{e^{hc/k\lambda T} - 1}$$

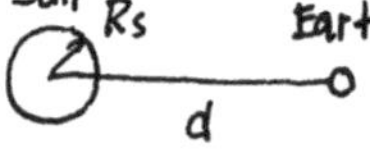

Take $\lambda = 0.5\,\mu m$ $\rightarrow$ $G_{\lambda\perp} = 1.79 \cdot 10^9\ W/(m^2 \cdot m)$
$= 1.79 \cdot 10^3\ W/(m^2 \cdot \mu m)$

Compare to curve in Figure 42.

■

Integration of the spectral entropy intensity of solar radiation according to the WRC spectrum (Figure 43) yields a value of 4620 W/(K · m² · sr), while the integral value of the energy intensity is $2.011 \cdot 10^7$ W/(m² · sr). Derive the equivalent blackbody temperature and calculate the entropy current density for such radiation near the Earth.

SOLUTION: Equations (140) and (141) yield

$$j_{E\perp} = \Omega_s \frac{c}{4\pi} a T^4 \qquad \text{(see also Problem 30)}$$

$$= \Omega_s \frac{3}{4} T j_{SE}$$

$$\Rightarrow i_{Eb} = \frac{3}{4} T i_{sb}$$

$$\rightarrow T = \frac{4}{3}\frac{i_{Eb}}{i_{sb}} = \frac{4}{3}\frac{2.011 \cdot 10^7}{4620} \text{ K} = 5800 \text{ K}$$

Entropy current density near the earth:

$$j_s = \Omega \cdot i_{sb} = \pi R_s^2/d^2 \; 4620 = 0.32 \text{ W/(K}\cdot\text{m}^2)$$

■

Consider the absorptance of cavities and rooms. Light falls from the outside on the opening of a cavity (which might be a room with a window for the opening). The surface area of the opening is A_a, while the area of the inner surfaces is A_i. The absorptance of the inner walls is assumed to be α_i (independent of the angle of incidence and the wavelength). a) Show that the total absorptance is given by

$$\alpha = \alpha_i\left[\alpha_i + (1-\alpha_i)\frac{A_a}{A_i}\right]^{-1}$$

if the opening is not covered. (*Hint:* Consider rays bouncing off the interior walls and assume that after each reflection there is a probability of A_a/A_i for the ray to escape through the hole.) b) Show that the result must be

$$\alpha = \tau\cdot\alpha_i\left[\alpha_i + (1-\alpha_i)\tau_d\frac{A_a}{A_i}\right]^{-1}$$

if there is a window with a transmittance to direct light τ and a transmittance to diffuse reflected light from the interior of τ_d.

SOLUTION:

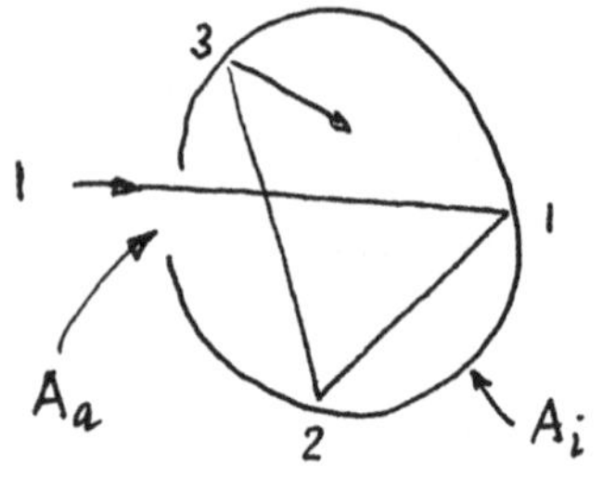

Consider a ray of unit intensity penetrating the opening. It bounces off the interior wall (where it is partly absorbed), and a fraction equal to the probability of hitting the opening will escape after each hit.

hit	bounced off wall	remains in cavity	escapes
1	$(1-\alpha_i)$	$(1-\alpha_i)(1-A_a/A_i)$	$(1-\alpha_i)A_a/A_i$
2	$(1-\alpha_i)^2(1-A_a/A_i)$	$(1-\alpha_i)^2(1-A_a/A_i)^2$	$(1-\alpha_i)^2(1-A_a/A_i)A_a/A_i$
3	$(1-\alpha_i)^3(1-A_a/A_i)^2$	$(1-\alpha_i)^3(1-A_a/A_i)^3$	$(1-\alpha_i)^3(1-A_a/A_i)^2A_a/A_i$
$\vdots$	$\vdots$	$\vdots$	$\vdots$

The heading of the third column of the previous table should actually read: remains in cavity as radiation; part of the radiation is absorbed by the walls and also remains inside. Therefore, we first have to calculate the reflectance of the cavity by adding up all the entries in the fourth column. From the reflectance we compute the absorptance.

$$\rho = (1-\alpha_i) A_a/A_i + (1-\alpha_i)^2 A/A_i (1-A_a/A_i) + (1-\alpha_i)^3 A_a/A_i(1-A_a/A_i) + \ldots$$

$$= (1-\alpha_i)\frac{A_a}{A_i}\left\{1 + (1-\alpha_i)\left(1-\frac{A_a}{A_i}\right) + (1-\alpha_i)^2\left(1-\frac{A_a}{A_i}\right)^2 + \ldots\right\}$$

$$= \frac{(1-\alpha_i) A_a/A_i}{1-(1-\alpha_i)(1-A_a/A_i)}$$

$$\Rightarrow \alpha = 1-\rho = \frac{1-(1-A_a/A_i)(1-\alpha_i) - (1-\alpha_i) A_a/A_i}{1-(1-\alpha_i) + (1-\alpha_i) A_a/A_i}$$

$$= \frac{1-(1-\alpha_i) + (1-\alpha_i) A_a/A_i - (1-\alpha_i) A_a/A_i}{\alpha_i + (1-\alpha_i) A_a/A_i}$$

$$= \frac{\alpha_i}{\alpha_i + (1-\alpha_i) A_a/A_i}$$

The second result is derived in an analogous fashion. ∎

For the collector and the situation discussed in Example 44, estimate the rate of production of entropy due to heat loss to the ambient if the energy current due to this effect is 40% of the irradiance. Compare this value to the rate of production of entropy due to absorption by the collector.

SOLUTION: From Example 44 we can describe the condition of the collector as follows:

optical efficiency: 0.80
total irradiance: $G = 1000\ W/m^2$
collector temperature: $T_c = 70°C$
ambient temperature: $T_a = 20°C$

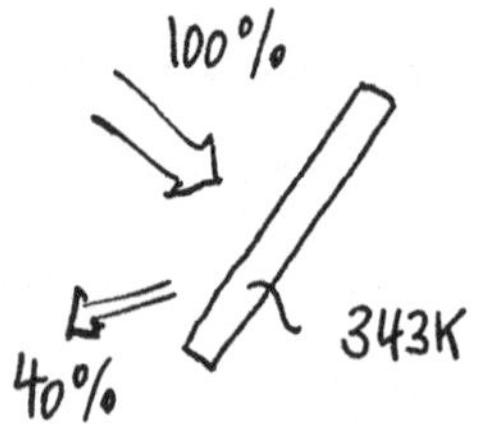

$$\pi_{S,\,loss} = I_{E,loss}\left(1/T_a - 1/T_c\right)$$

$$= 400\ W\left(1/293 - 1/343\right)1/K = 0.20\ W/K$$

$$\pi_{S,\,absorption} = \frac{1}{T_c}\,0.80\,AG = 2.3\ W/K$$

Both values hold for an area of $1\ m^2$. Obviously, in the case of solar collectors, absorption of radiation is much more dissipative than heat loss. However, reducing dissipation due to absorption is much more difficult than trying to reduce entropy production due to heat loss. (We would have to use very high temperature absorbers.) ■

CHAPTER 4

Solutions of Selected Problems

Imagine a power plant at the mouth of a river flowing into the ocean which uses the osmotic pressure difference between sea water and fresh water. If the river is carrying 1000 m³/s of fresh water, how large could the power of an ideal plant be?

SOLUTION:

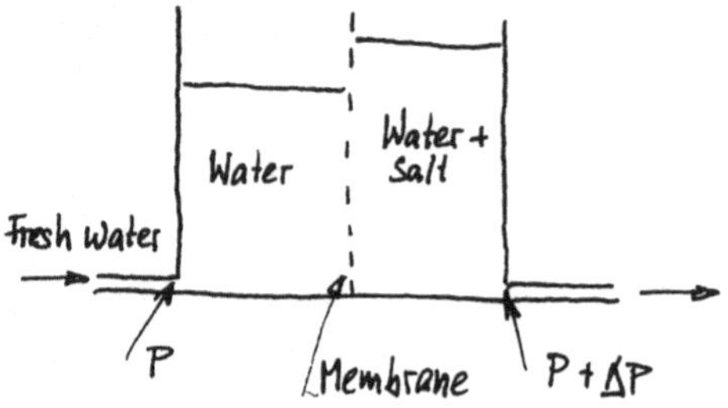

Turn the setup of Figure 2 into an "engine". It will work as derived below as long as the concentration of salt remains constant.

$$|\mathcal{P}| = |\Delta P \cdot I_V| \qquad |\Delta \bar{c}| = 1026 \text{ mole/m}^3 \text{ (Example 2)}$$
$$|\Delta P| = RT\,|\Delta \bar{c}| \qquad |\mathcal{P}| = 8.31 \cdot 300 \cdot 1026 \cdot 1000 \text{ W} = 2.56 \text{ GW}$$

Alternative solution: The "engine" effectively transfers pure solvent to a solution. If both are at the same (ambient) pressure, the chemical potential of the solvent in the solution is smaller:

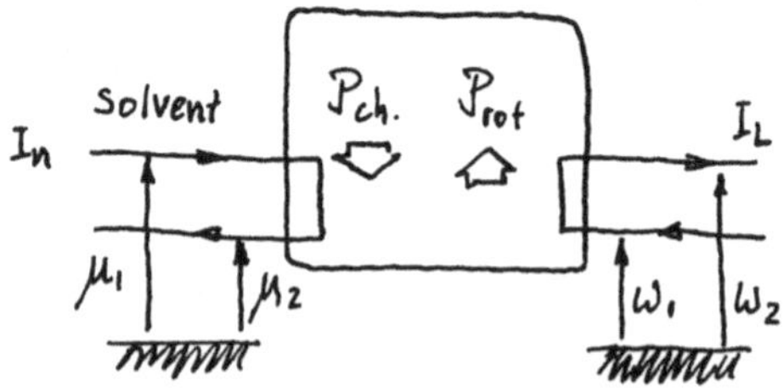

$$|\mathcal{P}_{chem}| = |\Delta\mu \, I_n|$$

$$\mu_2 = \mu_1 - RT \frac{\bar{v}_\ell}{\bar{v}_s} \quad \text{(Equation 98)}$$
$$= \mu_1 - RT \frac{n_s}{n_\ell}$$

$$|\mathcal{P}_{chem}| = 8.31 \cdot 300 \, \frac{2 \cdot 0.030/0.0585}{1.0/0.018} \, \frac{1}{0.018} \, 10^6 \text{ W} = 2.56 \cdot 10^9 \text{ W}$$

■

One mole of methane is burned with precisely the amount of air necessary for complete combustion. How much entropy and energy are given off by the combustion products if they are cooled from 800°C to 100°C?

SOLUTION: $CH_4 + 2O_2 + 7.44\,N_2 \rightarrow CO_2 + 2H_2O + 7.44\,N_2$

(See Table 1 on p. 459 for the amount of N_2 in air.)

We have a mixture of 1 mole CO_2, 2 moles H_2O, and 7.44 moles N_2 at 800°C which is cooled to 100°C. Let us treat the mixture as an ideal gas:

Entropy emitted: $S_e = \Delta S = (n_{CO2}\bar{C}_{p\,CO2} + n_{H2O}\bar{C}_{p\,H2O} + n_{N2}\bar{C}_{p\,N2}) \ln(T_2/T_1)$

Energy emitted: $W_e = \Delta E = (n_{CO2}\bar{C}_{V\,CO2} + n_{H2O}\bar{C}_{V\,H2O} + n_{N2}\bar{C}_{V\,N2})(T_2 - T_1)$

	$\bar{C}_p$ / J/(K·mole)	$\bar{C}_V$ / J/(K·mole)
H_2O	8.31 · 4.5	8.31 · 3.5
N_2	8.31 · 3.7	8.31 · 2.7
CO_2	8.31 · 5.2	8.31 · 4.2

Average values of the temperature coefficient of energy of the constituents of the gas have been taken from Figure 9 in Chapter 2 (page 189).

$$S_e = 8.31(5.2 + 2\cdot 4.5 + 7.44\cdot 3.7)\ln(373/1073)\ \text{J/K} = -370\ \text{J/K}$$

$$W_e = 8.31(4.2 + 2\cdot 3.5 + 7.44\cdot 2.7)(373 - 1073)\ \text{J/K} = -180\ \text{kJ}$$

■

Calcite ($CaCO_3$) decomposes to lime (CaO) and CO_2 at a temperature of 1195 K. Show how to obtain an estimate of this value from chemical properties of the substances at 25°C listed in the tables of the Appendix.

SOLUTION:

Table A.13

	μ/G at 298 K	α_μ G/K	μ/G at 1195 K
$CaCO_3$	−1129000	−92.9	$-1.12 \cdot 10^6$
CaO	−604000	−39.75	$-0.640 \cdot 10^6$
CO_2	−394000	−213.7	$-0.586 \cdot 10^6$

$$CaCO_3 \rightarrow CaO + CO_2$$

Condition:

$$\mu_{CaCO_3}(T) = \mu_{CaO}(T) + \mu_{CO_2}(T)$$

$$\mu(T) \approx \mu(T_0) + \alpha_\mu (T - T_0)$$

$$\rightarrow \quad T - T_0 \approx 900\ K \quad \Rightarrow \quad T \approx 1200\ K$$

■

Calculate the chemical potential of H_2O in its gaseous, liquid, and solid forms each at temperatures of 200 K, 300 K, and 400 K. From the results deduce which of the forms should be stable at the three temperatures listed.

SOLUTION: We use data on H_2O found in Table A.13 for 298 K, and calculate estimates of the chemical potentials for the three phases for 200 K, 300 K, and 400 K.

	μ/kG 298 K	α_μ/G/K	μ/kG 200 K	μ/kG 300 K	μ/kG 400 K
gaseous	−228.6	−188.7	−210.1	−229	−247.8
liquid	−237.2	−69.9	−230.3	−237	−244.3
solid	−236.6	−44.8	−232.2	−236	−241.2

The lowest values of the chemical potential for each of the three temperatures are underlined. For example, liquid H_2O is the stable form at 300 K.

■

Show how a measurement of the vapor pressure curve and the change of the volume of the fluid upon vaporization can be used to derive the changes of entropy, enthalpy, and energy of the fluid for the phase change.

SOLUTION:

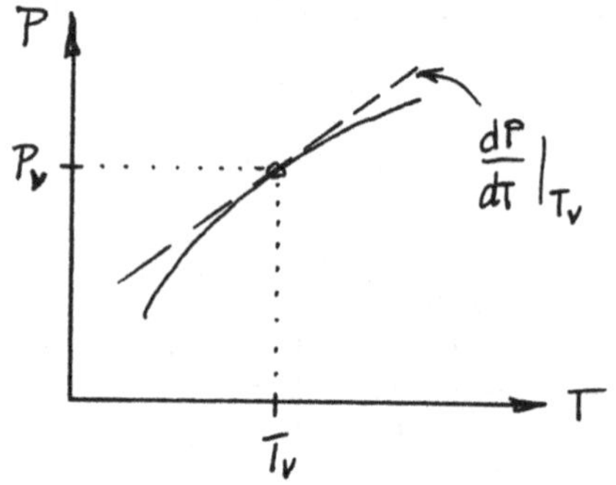

Measurement of the vapor pressure curve yields P_v and dP/dT at T_v.

Measurement of the change of volume ΔV yields

$$\bar{v}_g - \bar{v}_l = \frac{1}{n}\Delta V$$

Change of entropy:

$$\Delta S = n\,\Delta\bar{s} = n(\bar{v}_g - \bar{v}_l)\frac{dP}{dT}\Big|_{T_v} \quad \text{(Equation 92)}$$

$$= \Delta V \frac{dP}{dT}\Big|_{T_v}$$

Change of enthalpy:

$$\Delta H = n\,\Delta\bar{h} = n T_v \Delta\bar{s} \quad \text{(Equation 79)}$$

$$= \Delta V \cdot T_v \frac{dP}{dT}\Big|_{T_v}$$

Change of energy:

$$\Delta E = \Delta(H - PV) = \Delta H - P_v \Delta V \quad \text{(Equation 45)}$$

$$= \left(T_v \frac{dP}{dT}\Big|_{T_v} - P_v\right)\Delta V$$

You can use these relations for values for water found in Table A.16. ■

Two grams of CO_2 are dissolved in a liter of bottled water. Assuming that there only is carbon dioxide in the space above the water, how large is the pressure of the gas?

SOLUTION:

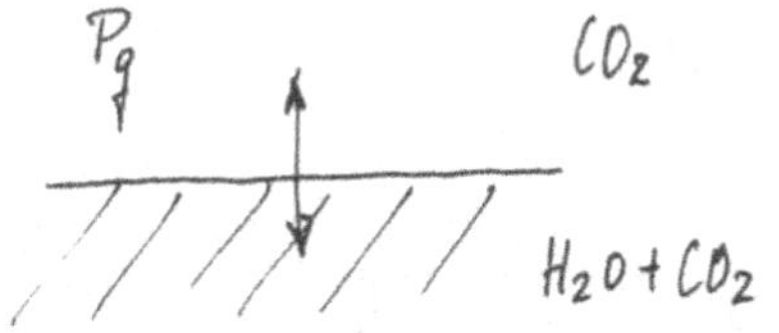

Chemical equilibrium between CO_2 in solution and CO_2 gas above the solution:

$$\mu_{CO_2}(gas) = \mu_{CO_2,s}$$

Equation (103): $\mu_g^\circ + RT \ln(P_g/P_0) = \mu_{aq}^\circ + RT \ln(\bar{c}_{aq}/\bar{c}_0)$

where P_0, μ_g°, μ_{aq}°, and $\bar{c}_0$ are values at standard conditions. From Table A.13 for CO_2:

$$\mu_g^\circ = -394.4 \text{ kG}$$
$$\mu_{aq}^\circ = -386.0 \text{ kG}$$

$$\rightarrow -394.40 \cdot 10^3 \text{ J/mole} + 8.31 \cdot 298 \text{ J/mole} \cdot \ln(P_g/P_0)$$
$$= -386.0 \cdot 10^3 \text{ J/mole} + 8.31 \cdot 298 \text{ J/mole} \cdot \ln\left(\frac{\frac{0.002}{0.044}}{1.0}\right)$$

$$\rightarrow \quad P_g/P_0 = 1.35 \quad \rightarrow \quad P_g = 1.35 \cdot 10^5 \text{ Pa}$$

■

Show that the relation between the mass of snow (modeled as ice) melting and the mass of water vapor condensing out of fog (see Problem 10) is given by

$$m_{snow} = 7.5 m_{vapor}$$

while the ratio of snow melting from condensation of fog and from cooling of rain from temperature T to T_{snow} is roughly $600/(T - T_{snow})$.[2]

SOLUTION:

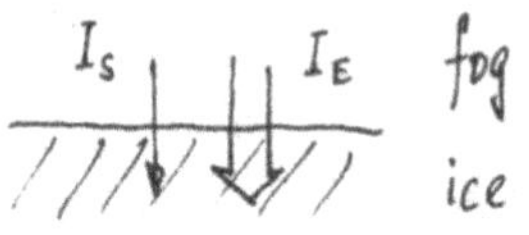

Energy exchanged:

$$Q/m_{fog} = h_{fog} - h_{water} = T_v (S_{fog} - S_{water})$$

$$\approx 2.5 \cdot 10^6 \text{ J/kg}$$

$(T_v = 273$ K; see Table A.16.)

Melting of ice: $Q/m_{ice} = 334 \cdot 10^3$ J/kg (Table A.4) $\rightarrow$ $m_{ice} = 7.5 \cdot m_{fog}$

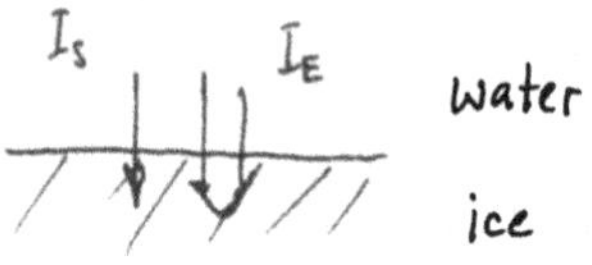

Cooling of water:

$$Q = m_{water} c_p (T - T_{ice})$$

$$\rightarrow \frac{m_{ice}}{m_{water}} = \frac{4200(T - T_{ice})}{3.34 \cdot 10^6} = 1.26 \cdot 10^{-2} (T - T_{ice})$$

Ratio of the two ways of melting ice:

$$\frac{m_{ice}(fog)}{m_{ice}(water)} = \frac{7.5}{1.26 \cdot 10^{-2} (T - T_{ice})} = \frac{600}{(T - T_{ice})}$$

■

See Bohren (1995), p. 79–81.

Determine the vapor pressure of Refrigerant 123 (R123) as a function of temperature. It is known that at 0°C the vapor pressure is 32.7 kPa. Use data from Figure 43 and compare your result with Figure 65.

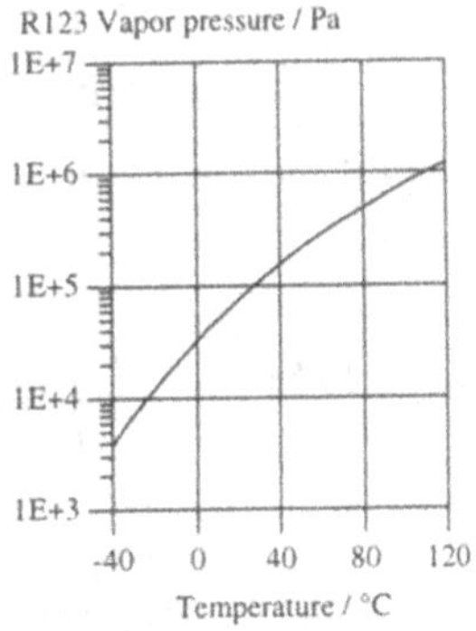

FIGURE 65. Problem 12.

$M_0 = 0.153$ kg/mole

SOLUTION:

Equation (92): $$\frac{dP}{dT} = \frac{\bar{s}_g - \bar{s}_l}{\bar{v}_g - \bar{v}_l}$$

ideal gas for g: $$\bar{v}_g - \bar{v}_l \approx \bar{v}_g = \frac{V_g}{n} = \frac{RT}{P}$$

$$\rightarrow \frac{dP}{dT} = \frac{P_v}{RT_v}(\bar{s}_g - \bar{s}_l) = \frac{P_v}{RT_v}\frac{1}{M_0}(s_g - s_l)$$

Values of P_v, T_v, s_g and s_l must be read from the figures, and the equation just derived must be integrated numerically. For intervals which are not too large:

$$\ln(P/P_0) = \frac{1}{R}\left(\frac{1}{T_0} - \frac{1}{T}\right)\Delta\bar{h}_v$$

This will be demonstrated for the interval from 0°C to 40°C. Take values at 0°C:

$$P_v(40°C) = P_v(0°C)\exp\left\{\frac{1}{R}\left(\frac{1}{T_0} - \frac{1}{T}\right) T_0 M_0 (s_g - s_l)\right\}$$

$$= 32.7\cdot 10^3\ \mathrm{Pa}\cdot \exp\left\{\frac{1}{8.31}\left(\frac{1}{273} - \frac{1}{313}\right) 273\cdot 0.153\ (800-125)\right\}$$

$$= 1.6\cdot 10^5\ \mathrm{Pa}$$

which corresponds to the value which can be read from Figure 65. ■

Show that Equation (98) can be written in the form

$$\mu_l(T,\rho_n) = \mu_l(T,\rho_{no}) - M_o RT \upsilon_l(\rho_n - \rho_{no})$$

where the indices l and s refer to the solvent (fluid) and the solute, respectively.

SOLUTION: Equation (98)

$$\mu_l(P_l) = \mu_l(P_o) - RTy$$

$$= \mu_l(P_o) - RT\frac{\bar{\upsilon}_l}{\bar{\upsilon}_s} \qquad \text{(Equation 97)}$$

$$= \mu_l(P_o) - RT\frac{n_s}{n_l + n_s} \approx \mu_l(P_o) - RT\frac{n - n_l}{n_l}$$

$$= \mu_l(P_o) - RT\frac{n/V - n_l/V}{n_l/V}$$

$$= \mu_l(P_o) - RT\frac{V}{n_l}(\rho_n - \rho_{no})$$

$$= \mu_l(P_o) - RT\bar{\upsilon}_l(\rho_n - \rho_{no})$$

$$= \mu_l(P_o) - M_o RT\upsilon_l(\rho_n - \rho_{no}).$$

■

Model the water turbine discussed in Section 4.4.1 as a cylinder having a piston capable of taking in and ejecting water at the desired pressures. Create a four-step cycle and display the processes in the μ-n and the P-V diagrams.

SOLUTION:

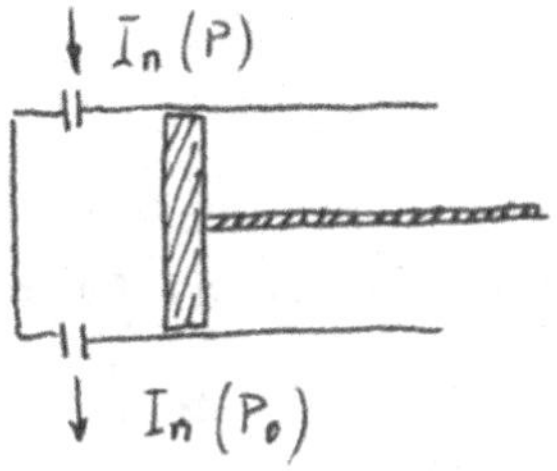

Step 1: Water is taken up by the empty cylinder at pressure P, and the piston moves to the right.

Step 2: Valves are closed, and the pressure of the water is decreased to P_0 (no change of volume of water).

Step 3: Water is ejected at pressure P_0.

Step 4: Valves are closed, and the pressure is increased to P (at $V = 0$).

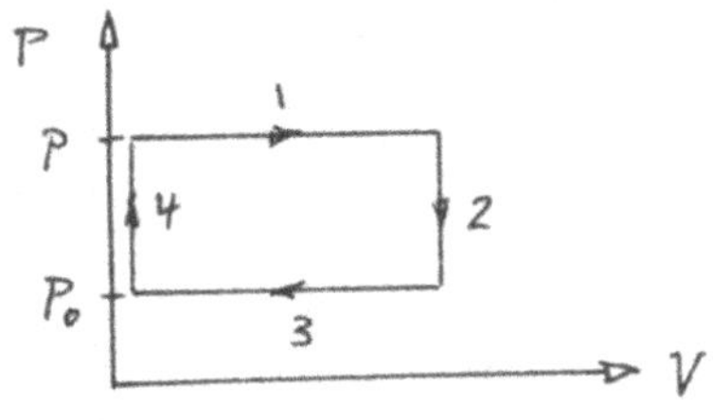

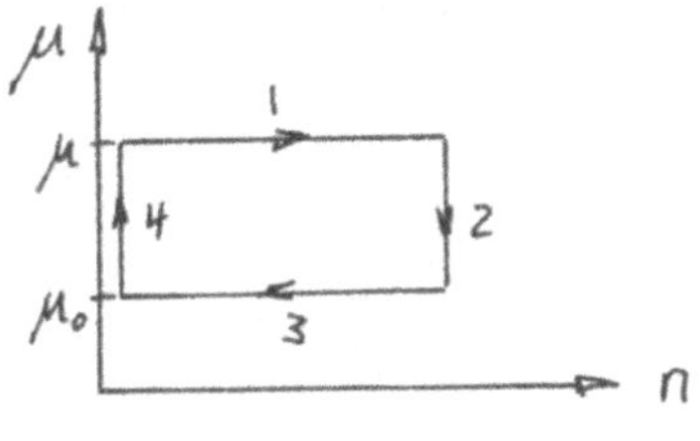

$$(P - P_0)\,\Delta V = (\mu - \mu_0)\,\Delta n$$

which yields the result presented in Equation (122). ■

Show that the chemical potential of a fluid takes the form

$$\mu = M_o[u + Pv - Ts]$$

if it is expressed using specific quantities of energy, volume, and entropy. How does this formula change if you include the flow velocity?

SOLUTION: The expression for the chemical potential not including the effects of velocity is given by Equation (140):

$$\begin{aligned}\mu &= \bar{u} + P\bar{v} - T\bar{s} \\ &= M_0 u + P M_0 v - T M_0 s \\ &= M_0 [u + Pv - Ts].\end{aligned}$$

Velocity effects are included in Equation (143):

$$\begin{aligned}\mu &= \bar{u} + P\bar{v} - T\bar{s} - \tfrac{1}{2} M_0 v^2 \\ &= M_0 [u + Pv - Ts - \tfrac{1}{2} v^2].\end{aligned}$$

■

Change the water turbine discussed in Section 4.4.1 to a compressed air engine. Model the engine as a cylinder having a piston, and an inlet and an outlet both with valves. Air is taken up at high chemical potential (pressure) and released again at lower potential. The piston moves in such a manner as to allow for an isothermal cycle of the gas. a) Show that you get the expression for the chemical potential of the ideal gas as a function of pressure (for constant temperature) derived above. b) Why is it possible to get the result by equating the mechanical power of the engine to the net flow energy current expressed solely in terms of the chemical energy current, even though we must heat the gas for the cycle to be performed isothermally?

SOLUTION:

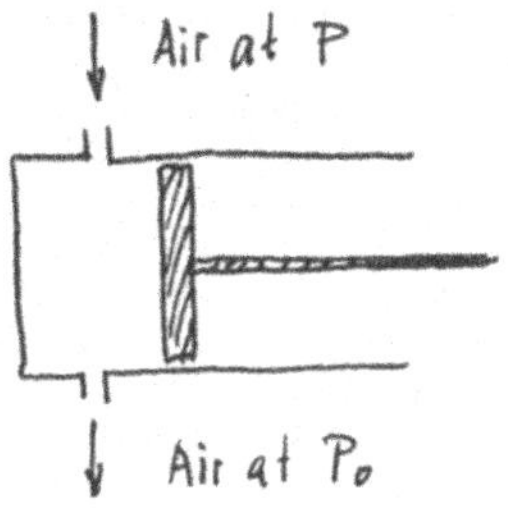

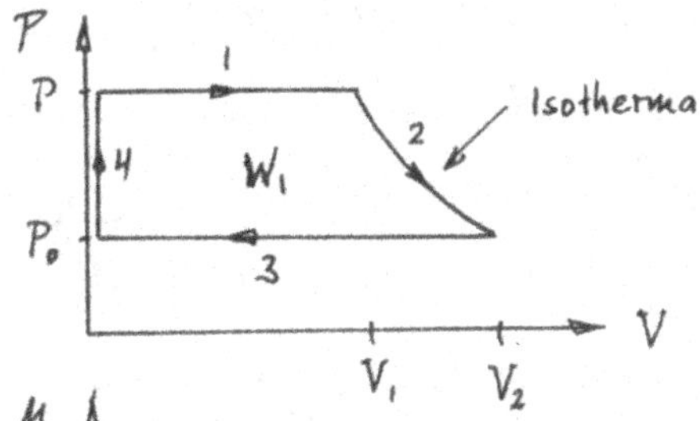

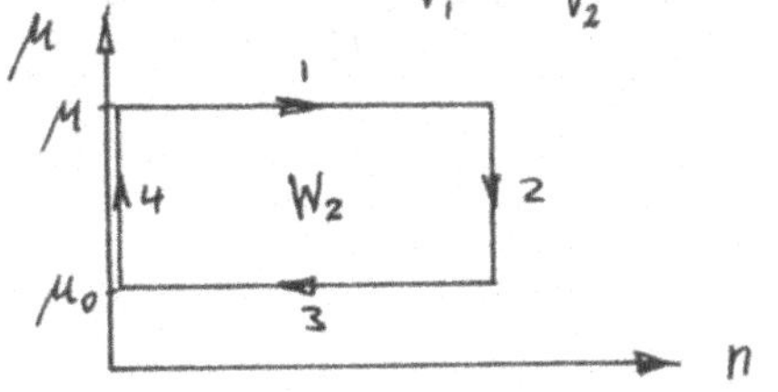

a) Both representations should yield the same result : $W_1 = W_2$

$$W_1 = P(V_1 - 0) + \int_{V_1}^{V_2} P\,dV - P_0(V_2 - 0)$$

$$= PV_1 - P_0V_2 + \int_{V_1}^{V_2} \frac{nRT}{V}\,dV$$

$$= \underbrace{PV_1 - P_0V_2}_{=0 \text{ since } T = const} + nRT \ln\left(\frac{V_2}{V_1}\right)$$

$$= nRT \ln(V_2/V_1) = nRT \ln(P/P_0)$$

$$W_2 = (\mu - \mu_0)\, n$$

$$\rightarrow \quad \mu - \mu_0 = RT \ln\left(\frac{P}{P_0}\right) \qquad \text{(see Equation 150)}$$

b) According to Section 4.4.5 (p. 517-518), the compressor has to be cooled. In the derivation presented in a, there was no mention of thermal processes. So the question arises if the result can be correct. We still get the right answer without considering the thermal side, since the balance of entropy demonstrates that thermal effects drop out of the balance of energy (see Equation 147).

■

Apply a control volume analysis to the determination of the temperature coefficient of enthalpy of air proposed in Example 19 of Chapter 2.

SOLUTION:

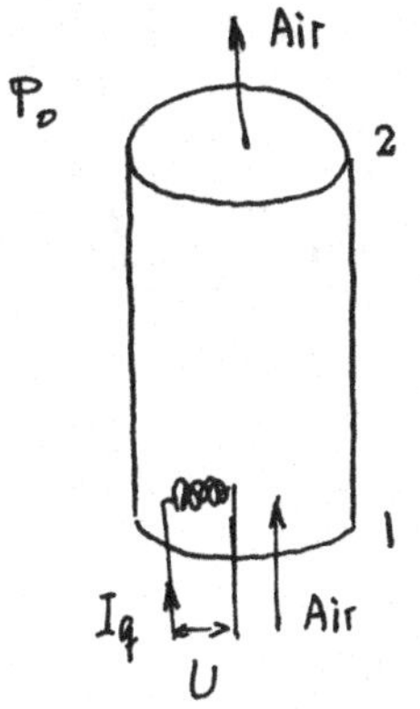

Steady-state balance of energy for open systems:

$$I_{E,conv,1} + I_{E,conv,2} + I_{E,el.} = 0$$

$$-c_p |I_m| (T_2 - T_1) + U I_q = 0$$

$$\rightarrow \quad c_p = \frac{U I_q}{|I_m| (T_2 - T_1)}$$

at Point 2: $|I_m| = \rho_2 |I_{V2}| = \rho_2 v A$

$$= \frac{M_0 P_0}{R T_2} v A$$

$$\rightarrow \quad c_p = \frac{R T_0 \cdot U I_q}{M_0 P_0 v A (T_2 - T_1)}$$

which is the same result as the one derived before in Example 19 of Chapter 2.

■

Hydrogen is burned with oxygen in a rocket engine. Knowing that the exit speed of the gas is around 3000 m/s, estimate the temperature of the stream.

SOLUTION:

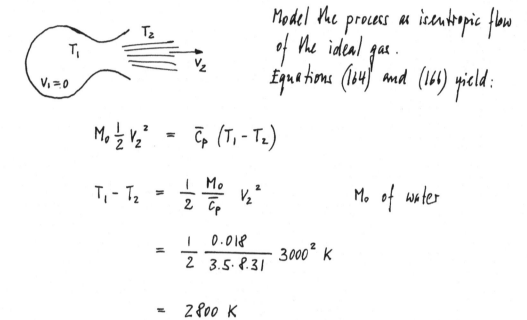

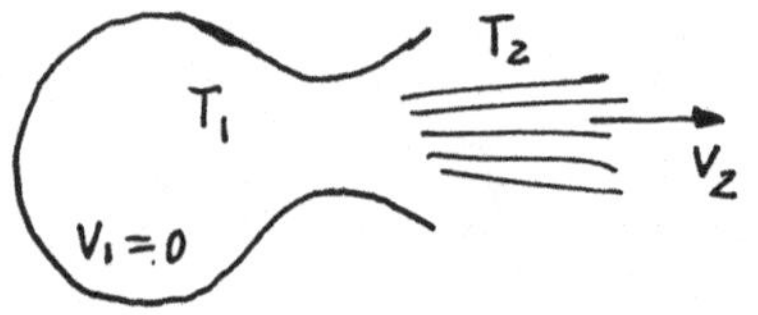

Model the process as isentropic flow of the ideal gas.
Equations (164) and (166) yield:

$$M_o \frac{1}{2} V_2^2 = \bar{C}_p (T_1 - T_2)$$

$$T_1 - T_2 = \frac{1}{2} \frac{M_o}{\bar{C}_p} V_2^2 \qquad M_o \text{ of water}$$

$$= \frac{1}{2} \frac{0.018}{3.5 \cdot 8.31} 3000^2 \text{ K}$$

$$= 2800 \text{ K}$$

If we assume the exit temperature to be around 1000 K, the temperature inside the engine will approach 4000 K. ■

Use the equation for the pressure as a function of height above ground in an isothermal atmosphere to derive the chemical potential of the ideal gas at constant temperature.

SOLUTION:

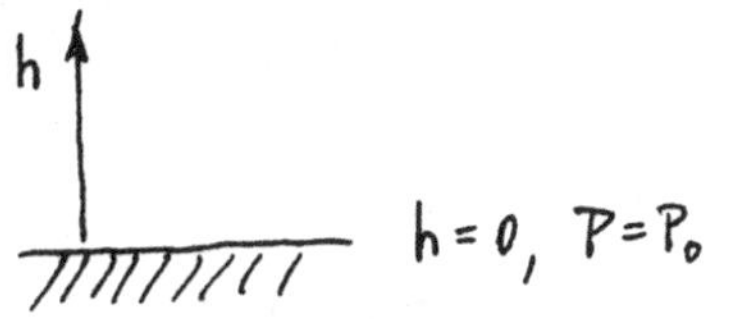

The law of hydrostatic equilibrium for the atmosphere:

$$\frac{dP}{dh} = -\rho g$$

Air as an ideal gas: $\rho = \frac{M_0 P}{RT} \rightarrow \frac{dP}{dh} = -\frac{M_0 g}{RT} h$ $\quad T = \text{const.}$

Solution of the equation of HE: $\int_{P_0}^{P} \frac{dP}{P} = -\frac{M_0 g}{RT} \int_0^h dh$

$$\rightarrow \ln(P/P_0) = -\frac{M_0 g}{RT} h$$

Equation (181): $\Delta\mu = -M_0 \Delta\varphi_G = -M_0 g h$

$$\Rightarrow \Delta\mu = RT \ln(P/P_0)$$

■

Derive the chemical potential for an incompressible fluid as a function of temperature at constant pressure.

SOLUTION: The derivation is analogous to the one performed for the ideal gas, with the exception that for incompressible fluids $c_p = c_v = c$.

Equation (53):

$$\mu(T,P) = \mu(T_0,P) + \bar{h} - \bar{h}_0 - (T\bar{s} - T_0\bar{s}_0)$$

$$\bar{h} - \bar{h}_0 = \bar{c}(T - T_0) \quad \text{if } \bar{c} = \text{const.}$$

$$\bar{s} - \bar{s}_0 = \bar{c}\ln(T/T_0)$$

$$\Rightarrow \mu(T,P) = \mu(T_0,P) + \bar{c}(T-T_0) - \left(T\left(\bar{s}_0 + \bar{c}\ln(T/T_0)\right) - T_0\bar{s}_0\right)$$

$$= \mu(T_0,P) + \bar{c}(T-T_0) - T\bar{c}\ln(T/T_0) - (T-T_0)\bar{s}_0$$

■

Consider the following strongly simplified model of the accretion of a planet. Matter with a temperature T_o falls from far away onto the surface of a growing planet. (The planet is surrounded by a gas of temperature T_o.) Assume the surface of the planet to radiate as a gray body. a) If the effect of the rate of change of the temperature can be neglected, show that the surface temperature at an instantaneous value of the radius can be calculated using

$$|I_m|c_p(T - T_o) + e4\pi r^2\sigma(T^4 - T_o^4) = |I_m|G\frac{m(r)}{r}$$

where I_m is the flux of mass falling upon the planet, c_p is its specific temperature coefficient of enthalpy, and e is the emissivity of the surface. (*Hint:* Treat the surface as an open control volume and consider the law of balance of energy for this system; first derive the instationary model.) b) Show that this is equivalent to

$$\rho c_p(T - T_o)\frac{dr}{dt} + e\sigma(T^4 - T_o^4) = \rho G\frac{m(r)}{r}\frac{dr}{dt}$$

c) Model accretion such that the rate of change of the radius of the planet is given by[3]

$$\frac{dr}{dt} = k_1 t^2 \sin(k_2 t)$$

At $t = 0$ and at $t = t_a$ (total accretion time) this function is supposed to vanish, and the radius of the planet grows from 0 to R during this period. Show that this leads to the following expressions for the constants k_1 and k_2:

$$k_1 = \frac{R}{t_a^3(1/\pi - 4/\pi^3)}$$

$$k_2 = \frac{\pi}{t_a}$$

d) Calculate $T(r)$ for the following values of the parameters. $T_o = 100$ K, $\rho = 5500$ kg/m³, $c_p = 800$ J/(K · kg), $R = 6.4 \cdot 10^6$ m, $t_a = 5 \cdot 10^5$ years, $e = 1$. (You should get the largest temperature, roughly 1000 K, at a radius of 5000 km.)

SOLUTION:

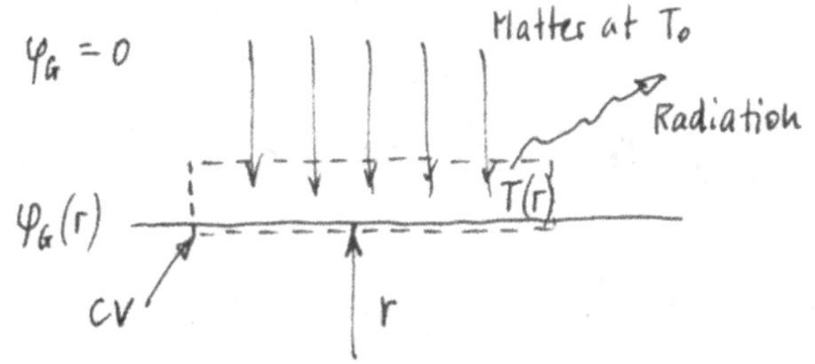

Anderson (1989), p. 3.

a) Balance of energy for control volume :

$$\dot{E}_{cv} + I_{E,conv} + I_{E,rad} = \Sigma_{E,grav.}$$

$I_{E,conv} = -c_p |I_m| T_0$ convective current due to infalling matter

$I_{E,rad} = A(r) e \sigma (T(r)^4 - T_0^4)$ radiative current

$\Sigma_{E,grav} = \dot{m} \varphi_G(r)$ gravitational source rate due to matter appearing at r

$\dot{m} = |I_m|$ balance of mass

$\dot{E} = m\dot{e} + \dot{m}e \; , \; e = c_p T$ capacitive" law

$$\Rightarrow \; m\dot{e} + \dot{m} c_p T - c_p |I_m| T_0 + 4\pi r^2 \varepsilon \sigma (T(r)^4 - T_0^4) = \dot{m} \frac{Gm}{r}$$

Steady-state (e does not change at a location with time):

$$|I_m| c_p (T - T_0) + 4\pi r^2 \varepsilon \sigma (T(r)^4 - T_0^4) = |I_m| G \cdot m / r$$

b) $\dot{m} = 4\pi r^2 \rho \frac{dr}{dt} \; \Rightarrow \; |I_m| = 4\pi \cdot r^2 \rho \frac{dr}{dt}$

$$\Rightarrow \; 4\pi r^2 \rho \frac{dr}{dt} c_p (T - T_0) + 4\pi r^2 \varepsilon \sigma (T(r)^4 - T_0^4) = 4\pi r^2 \rho \frac{dr}{dt} \frac{Gm}{r}$$

$$\Rightarrow \; \rho c_p (T - T_0) \frac{dr}{dt} + \varepsilon \sigma (T(r)^4 - T_0^4) = \rho G \frac{m(r)}{r} \frac{dr}{dt}$$

c) Assume $\frac{dr}{dt} = k_1 t^2 \sin(k_2 t)$ t_a: accretion time
R: final radius of planet

Conditions: 1. $\left.\frac{dr}{dt}\right|_{t=t_a} = 0 \rightarrow \sin(k_2 t_a) = 0$
$\rightarrow k_2 t_a = \pi$

2. $R = \int_0^{t_a} k_1 t^2 \sin(k_2 t)\, dt$

$$= k_1 \left[\frac{2t}{k_2^2} \sin(k_2 t) - \left(\frac{t^2}{k_2} - \frac{2}{k_2^3} \right) \cos(k_2 t) \right]_0^{t_a}$$

$$= k_1 \left[\frac{2t_a}{k_2^2} \sin(k_2 t_a) - \left\{ \left(\frac{t_a^2}{k_2} - \frac{2}{k_2^3} \right) \cos(k_2 t_a) + \frac{2}{k_2^3} \right\} \right]$$

With 1: $R = k_1 \left(\frac{t_a^2}{k_2} - \frac{4}{k_2^3} \right) = k_1 \left(\frac{1}{\pi} - \frac{4}{\pi^3} \right) t_a^3 \Rightarrow k_1 = \frac{R}{t_a^3 (1/\pi - 4/\pi^3)}$

d)

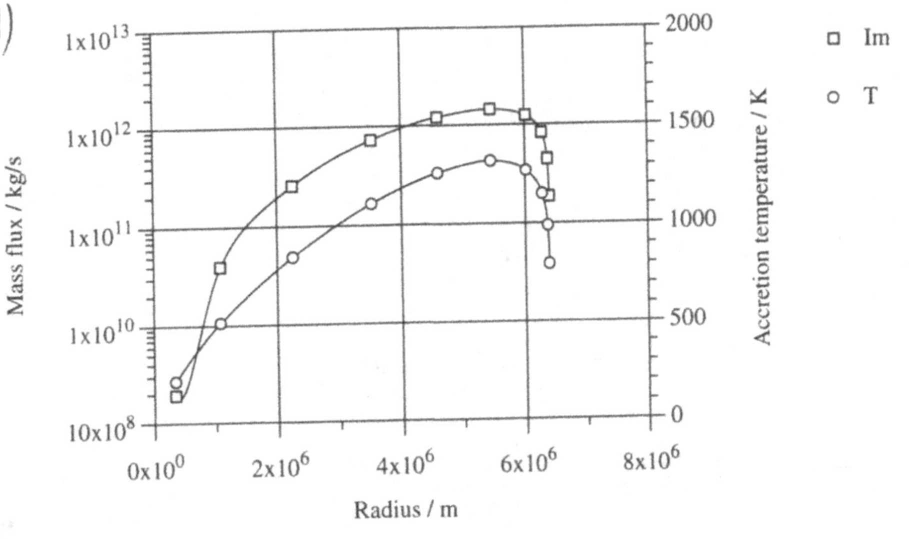

Model the heating of the air in a room as follows. As entropy is added to the air it expands and diffuses through the walls such as to leave the pressure in the room at a constant value. Now take a constant value for the heating power. Show that in this case the temperature of the air rises according to

$$T(t) = T_o \exp\left(\frac{\mathcal{P}_{heating}\, t}{7/2\, PV}\right)$$

if a value of $7/2R$ is taken for the molar temperature coefficient of enthalpy of air. Note that it is assumed that the air remaining in the room does not lose any heat to the surroundings.

SOLUTION:

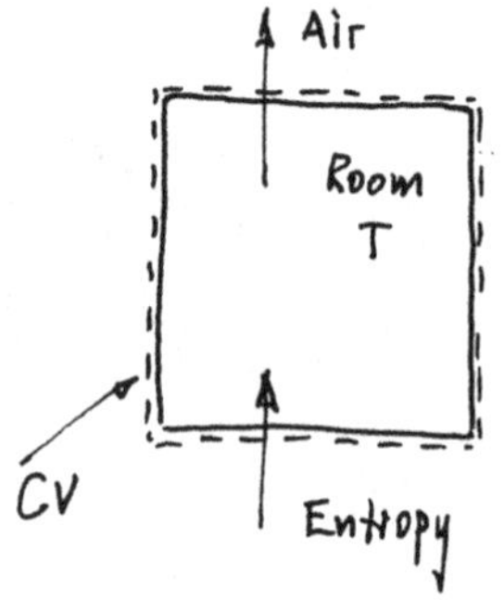

Balance of entropy:

$$\dot{S}_{cv} = \frac{|I_{E,th}|}{T} - s|I_m|$$

$$\dot{S}_{cv} = m\dot{s}_{cv} + \dot{m}s_{cv} \quad \text{where } \dot{s}_{cv} = \frac{c_p}{T}\dot{T}$$

Balance of mass: $\dot{m} = -|I_m|$

$$\Rightarrow \left.\begin{aligned} m\frac{c_p}{T}\dot{T} &= \frac{I_{E,th}}{T} \\ m = \rho V &= \frac{M_o P}{RT}V \end{aligned}\right\} \quad c_p \frac{M_o PV}{RT^2}\dot{T} = \frac{|I_{E,th}|}{T}$$

$$\frac{M_o c_p}{R} = \frac{7}{2} \Rightarrow \frac{7}{2}PV\frac{\dot{T}}{T} = |I_{E,th}|$$

$$\Rightarrow \int_{T_o}^{T}\frac{dT}{T} = \frac{|I_{E,th}|}{7/2\, PV}\int_0^t dt$$

$$\Rightarrow \ln(T/T_o) = \frac{|I_{E,th}|}{7/2\, PV}\cdot t$$

■

Express the efficiency of the Rankine cycle in terms of the average temperature of heating.

SOLUTION:

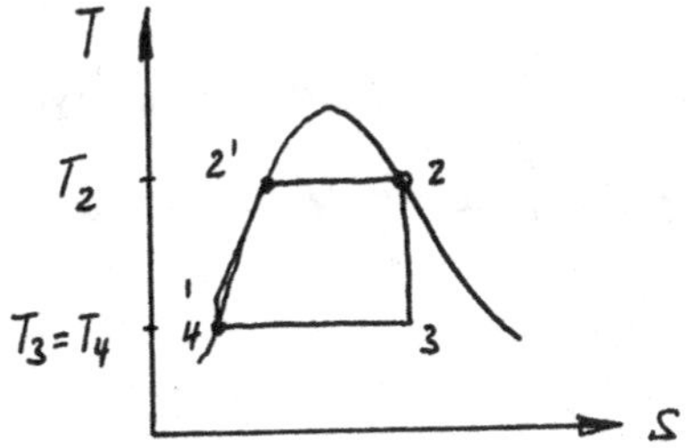

Efficiency $\eta = \dfrac{|I_{E,23}|}{|I_{E,12}|} = \dfrac{|W_{23}|}{|Q_{12}|}$

where we neglect the energy necessary for compressing the liquid $(4 \to 1)$.

$$|Q_{12}| = \int_1^2 T\dot{S}\,dt = \int_1^2 T\,dS$$

$$\Rightarrow \quad |Q_{12}| = \bar{T}(S_2 - S_1)$$

$\bar{T}$: average temperature during heating

$$|W_{23}| = |Q_{12}| - |Q_{34}|$$

$$= \bar{T}(S_2 - S_1) - T_3(S_3 - S_4)$$

Since $S_2 - S_1 = S_3 - S_4$ we have:

$$\eta = \frac{\bar{T}(S_2 - S_1) - T_3(S_3 - S_4)}{\bar{T}(S_2 - S_1)} = 1 - T_3/\bar{T}$$

■

Discuss the effect of changing the upper and the lower operating pressures of the Rankine cycle. Why is a condenser used in vapor power plants if steam leaving the turbine could be discharged directly to the environment?

SOLUTION:

Consider a Rankine cycle as in Figure 50 (or in Problem 30). Remember the approximate expression for the efficiency of such a cycle in terms of temperatures (Problem 30):

$$\eta = 1 - T_3/\bar{T}$$

where $\bar{T}$ is the average temperature during heating.

Now increase the upper operating pressure. This will increase T_2, leading to larger $\bar{T}$, and therefore to increased efficiency. Likewise, decreasing the lower operating pressure decreases T_3, which again leads to an increased efficiency.

Discharging steam directly to the environment means that the lowest possible pressure is 1 bar, corresponding to $T_3 = 100°C$. If a condenser is included, lower condensing pressures can be attained which increases the efficiency of the plant.

■

Assume the furnace of the Carnot cycle proposed in Problem 32 to operate at 850°C and the condenser at 20°C, respectively. The cycle undergone by the working fluid is supposed to be the same as before. Calculate the rate of production of entropy and the rate of loss of availability.

SOLUTION:

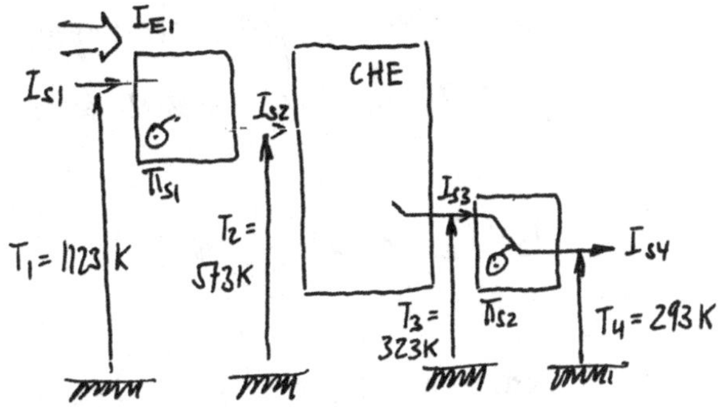

In Problem 32, a Carnot cycle runs between 300°C and 50°C, respectively. Here, we model the entire process as an endoreversible engine (see Chapter 1.7).

$$\Pi_s = |I_{s2}| - |I_{s1}| + |I_{s4}| - |I_{s3}|$$

$$= |I_{E1}| \left(\frac{1}{T_2} - \frac{1}{T_1}\right) + T_3 |I_{s3}| \left(\frac{1}{T_4} - \frac{1}{T_3}\right)$$

$$= |I_{E1}| \left(\frac{1}{T_2} - \frac{1}{T_1}\right) + T_3 \frac{|I_{E1}|}{T_2} \left(\frac{1}{T_4} - \frac{1}{T_3}\right)$$

$$= |I_{E1}| \left(\frac{T_3}{T_2 T_4} - \frac{1}{T_1}\right) \quad \Rightarrow \quad \Pi_s = |I_{E1}| \cdot 1.0 \cdot 10^{-3} \text{ 1/K}$$

$$\mathcal{L} = T_4 \Pi_s = |I_{E1}| \left(\frac{T_3}{T_2} - \frac{T_4}{T_1}\right) = \left(\frac{323}{573} - \frac{293}{1123}\right) |I_{E1}|$$

$$\mathcal{L} = 0.30 \, |I_{E1}|$$

■

a) Estimate the efficiency of a vapor power cycle without superheating designed for the fluid R123. The heat is supposed to be delivered by solar collectors such as vacuum tubes. Saturated liquid enters the evaporator at a pressure of 8.0 bar, while the condenser operates at a temperature of 30°C. (Use property data found in Figure 43 and Figure 65.) b) If the collectors deliver an energy current of 350 W per square meter of collector area, what is the minimum collector area needed per kW of power of the engine?[4]

SOLUTION:

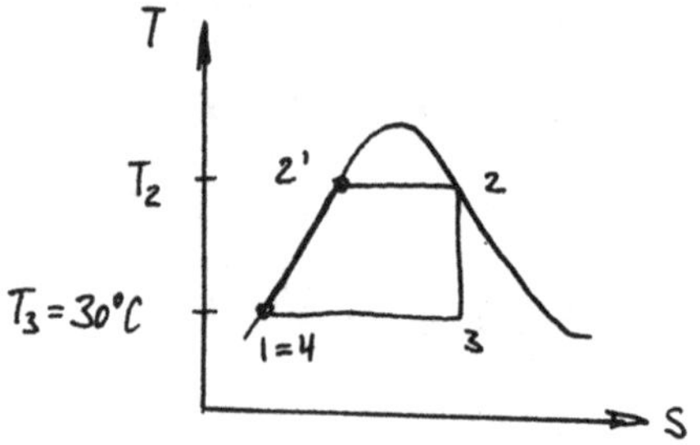

a) Determine T_2 from the vapor pressure of R 123 (Figure 65):

$$T_2 (P_2 = 8\,bar) = 100°C$$

Determination of the efficiency (see p. 544 - 545)

$$\eta \approx 1 - \frac{|I_{E\,34}|}{|I_{E\,12}|} = 1 - \frac{T_3 (S_3 - S_4)}{h_{2'} - h_1 + T_2 (S_2 - S_{2'})}$$

Determination of property data:

Figure 43: $S_2 = 0.80\ kJ/(K \cdot kg)$ $\quad S_{2'} = 0.45\ kJ/(K \cdot kg)$

$S_3 = S_2 = 0.80\ kJ/(K \cdot kg)$ $\quad S_4 = 0.24\ kJ/(K \cdot kg)$

Figure 65: $P_3 (T_3) = 1.1 \cdot 10^5\ Pa$

Figure 43: $h_{2'} (P_2) = 140\ kJ/kg$ $\quad h_1 (P_3) = 70\ kJ/kg$

See Koch (1993) for a discussion of the merits of different types of fluids used in such a scheme with or without superheating.

$$\Rightarrow \quad \eta = 1 - \frac{303\,(800-240)}{140\cdot 10^3 - 70\cdot 10^3 + 373\,(800-450)} = 0.15$$

Remember that this is the thermal efficiency of an ideal fluid running through a Rankine cycle. The real efficiency of the engine would be somewhat smaller (say 0.10).

b) $$\mathcal{P} = \eta\,|I_{E,in}| = \eta\, A\,\frac{|I_{E,in}|}{A}$$

$$\rightarrow \quad A = \frac{\mathcal{P}}{\eta\,|I_{E,in}|/A} = \frac{1000}{0.15\cdot 350}\ \mathrm{m^2} = 19\ \mathrm{m^2}$$

■

Estimate the amount of entropy produced in the throttling process of a refrigerant if the following data are given: the initial and the final pressure, and the initial and the final specific volume of the fluid.

SOLUTION:

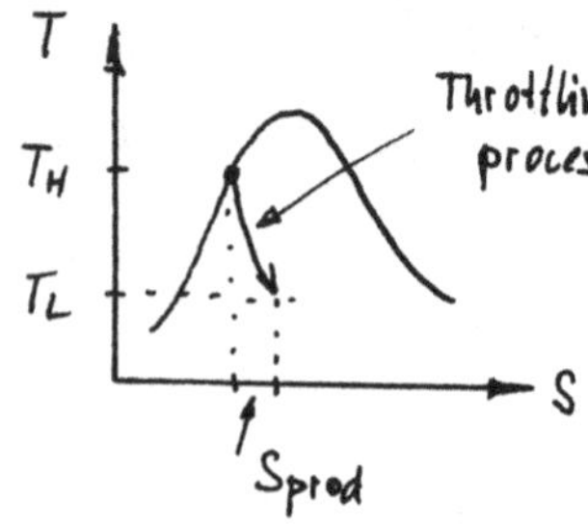

The given data are sufficient to perform a detailed calculation of the process if property data are available.
Lacking this information, we still can perform a very rough estimate. Entropy production is the result of dissipation of the energy released by the fluid flowing through the given pressure difference: $s_{prod} \approx \Delta P\, v_{av} / T_{av}$ where $\Delta P = P_H - P_L$, and $v_{av} \approx 0.5\, v_L$ since $v_H \ll v_L$, and $T_{av} \approx 0.5\,(T_H + T_L)$. This last value has to be estimated. For an example involving R123 calculated according to these rules, a value of s_{prod} is obtained which is roughly twice the actual value.

Note: With intermediate data on P and v (and associated temperatures), and summing over smaller intervals, the idea presented here yields the correct value. In general, there is a disproportionately large increase of volume for the lower pressure range; therefore, the estimate presented is too large. ■

If 60°C water is to be produced with solar collectors, why should they be operated in such a way that they deliver water of exactly 60°C?

SOLUTION: If we have solar collectors available, and if we require 60°C water, we have three options:

1. Produce the 60°C water using the collectors.
2. Let the collector operate with a smaller mass flux which will result in hotter water; we can later mix cold water with what we have from the collectors.
3. Produce water at lower temperature, and later heat it using electricity or fuels.

It turns out that Scheme 1 has the lowest rate of production of entropy per mass flux of 60°C water overall.

Scheme 1: Balance of energy:

$$c_p I_m (T_{out} - T_{in}) = F_R A \left((\tau\alpha) G - U_L (T_{in} - T_a) \right)$$

$$F_R = \frac{c_p I_m}{A U_L} \left(1 - \exp\left(-F' \frac{A U_L}{c_p I_m} \right) \right)$$

Balance of entropy:

$$\Pi_s = \frac{1}{T_a} \left(A (\tau\alpha) G - c_p I_m (T_{out} - T_{in}) \right) + c_p I_m \ln\left(\frac{T_{out}}{T_{in}} \right)$$

With $A = 10\,m^2$, $c_p = 4200\,J/(K \cdot kg)$, $G = 800\,W/m^2$, $(\tau\alpha) = 0.75$, $U_L = 4.0\,W/(Km^2)$, $F' = 0.90$, $T_{in} = T_a = 293\,K$, $T_{out} = 333\,K$ we get:

$$I_m = 0.0276 \text{ kg/s}$$
$$\Pi_s / I_m = 706 \text{ J/(K}\cdot\text{kg)}$$

Scheme 2: I_m will be smaller than for Scheme 1, and T_{out} will be larger than 333 K. If we mix a current I_{m2} of 293 K water with I_m from the collector, where

$$I_m T_{out} + I_{m2} T_a = (I_{m1} + I_{m2}) \cdot 333\text{K}$$

we have an additional rate of production of entropy as the result of mixing of the two fluid streams:

$$\Pi_{s2} = c_p (I_{m1} + I_{m2}) \ln\left(\frac{333\text{K}}{T_a}\right) - c_p I_m \ln\left(\frac{T_{out}}{T_a}\right)$$

The total rate of production per total mass flux $(\Pi_s + \Pi_{s2})/(I_m + I_{m2})$ is shown in the figure.

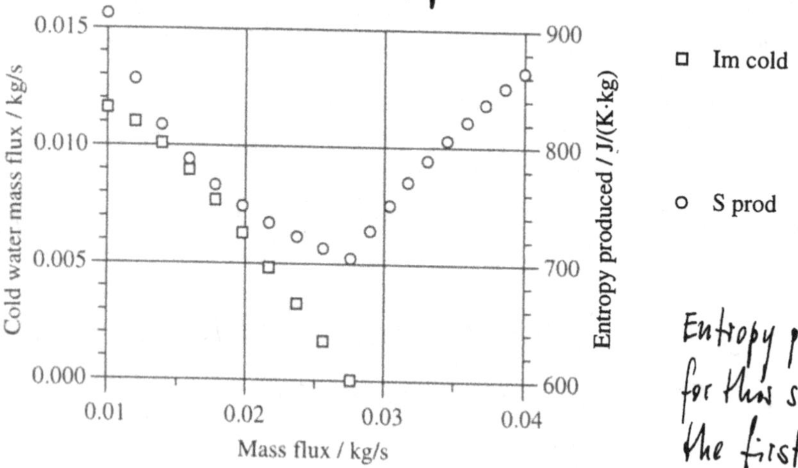

Entropy production is larger for this scheme than for the first.

Scheme 3: This scheme requires some interpretation. T_{out} will be less than 33K. If we simply add the entropy necessary to heat the water from T_{out} to 333K to the rate of production of entropy

$$\Pi_{s,tot} = \Pi_s + c_p I_m \ln\left(333K / T_{out}\right)$$

we get the confusing result that using electricity or fuels produces less entropy than solar heaters. While this is true if looked at in isolation, we should compare the production of warm water by any scheme with what would happen if we did not want any warm water at all. Entropy production would then be the result of absorption of solar radiation, which otherwise would end up in a collector, in the environment at T_a:

$$\Pi_{s,ref} = \frac{1}{T_a} A(\tau\alpha) G$$

Now, with solar heaters, Π_s is smaller than this value; compared to what nature does without us, producing hot water in solar heaters <u>decreases</u> irreversibility. Using electricity or fuels for heating produces entropy in addition to what would happen anyway. In the figure on the previous page, for $I_m > 0.0276$ kg/s, Π_s / I_m is shown where I have arbitrarily set $\Pi_s / I_m = 706$ J/K·kg at $T_{out} = 333K$.

■

A fluid is flowing through a pipe which is being heated from outside. (This setup can be found in line focus concentrators for solar radiation.) Show that the differential equation for the temperature of the fluid as a function of position in the direction of flow is given by

$$c_p I_m \frac{dT}{dx} = F'\left[\sigma_E - 2\pi r_o U_L (T - T_a)\right]$$

where

$$F' = \frac{1/U_L}{1/U_{fa}}$$

can be called the efficiency factor of the tubular heater. U_L is the heat loss coefficient from the pipe to the surroundings (see Problem 15 of Chapter 3 for a numerical example) while U_{fa}, which represents the heat transfer coefficient from the fluid to the environment, was calculated in Problem 15 of Chapter 3. σ_E is the rate of absorption of energy per length of pipe, and T_a is the ambient temperature.

SOLUTION:

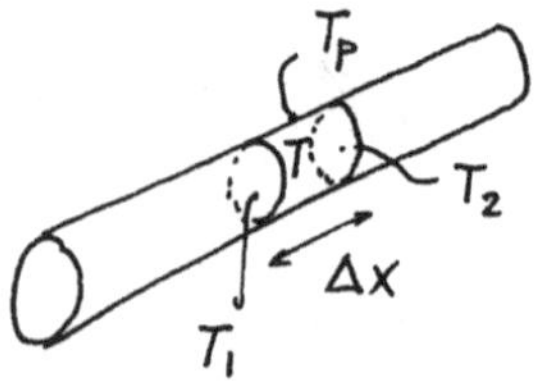

Balance of energy for segment of pipe:

(1) $0 = \Delta x \sigma_E - (T_p - T_a) U_L \Delta x 2\pi r$
$$- \Delta x 2\pi r U_{pf} (T_p - T)$$

where U_{pf} is the heat transfer coefficient from the pipe to the fluid.

Balance of energy for pipe plus fluid:

(2) $0 = \Delta x \sigma_E - U_L (T_p - T_a) \Delta x 2\pi r - c_p I_m (T_2 - T_1)$

Solve Eq. (1) and insert in (2):

$$2\pi r \Delta x \left(U_L + U_{pf}\right) T_p = \Delta x \sigma_E + 2\pi r \Delta x\, U_L T_a + 2\pi r \Delta x\, U_{pf} T$$

$$c_p I_m \left(T_2 - T_1\right) = \Delta x \sigma_E - 2\pi r \Delta x\, U_L \left(\frac{\Delta x \sigma_E + 2\pi r \Delta x U_L T_a + 2\pi r \Delta x\, U_{pf} T}{2\pi r\, \Delta x \left(U_L + U_{pf}\right)} - T_a \right)$$

$$= \Delta x\, \sigma_E - \frac{U_L}{U_L + U_{pf}} \left(\Delta x\, \sigma_E + 2\pi r\, \Delta x\, U_{pf} \left(T - T_a\right)\right)$$

$$= \Delta x\, \sigma_E \left(1 - \frac{U_L}{U_L + U_{pf}} \right) - 2\pi r\, \Delta x\, \frac{U_{pf}}{U_L + U_{pf}}\, U_L \left(T - T_a\right)$$

$$= \frac{U_{pf}}{U_L + U_{pf}} \left[\Delta x\, \sigma_E - 2\pi r\, \Delta x\, U_L \left(T - T_a\right)\right]$$

$$= F' \Delta x \left[\sigma_E - 2\pi r\, U_L \left(T - T_a\right)\right]$$

$$\Rightarrow \quad c_p I_m \frac{dT}{dx} = F' \left[\sigma_E - 2\pi r\, U_L \left(T - T_a\right)\right]$$

Note:
$$\frac{U_{pf}}{U_{pf} + U_L} = \frac{1}{1 + U_L / U_{pf}} = \frac{1/U_L}{1/U_L + 1/U_{pf}}$$

$$= \frac{1/U_L}{1/U_{fa}} = F'$$

■

Compare balanced counter-flow and parallel-flow heat exchangers, both with an *NTU* of 5. Which of the exchangers has the higher effectiveness?

SOLUTION: Relations for heat exchangers are given on p. 572-573.

Effectiveness of balanced counterflow heat exchanger:

Equation (222): $$\varepsilon_{cf} = \frac{NTU}{1+NTU} = \frac{5}{1+5} = 0.833$$

Effectiveness of balanced parallel flow heat exchanger:

Equation (223): $$\varepsilon_{pf} = \frac{1-\exp(-NTU(1+C^*))}{1+C^*}$$

where $C^* = \frac{(c_p \dot{m})_{min}}{(c_p \dot{m})_{max}}$, $C^* = 1$ for balanced heat exchanger

$$\Rightarrow \quad \varepsilon_{pf} = \frac{1-\exp(-5\cdot 2)}{1+1} = 0.500$$

■

Compare balanced and unbalanced counter-flow heat exchangers. If they are built identically, do they have the same effectiveness? (For concreteness, take the smaller of the two capacitance flow rates to be equal to the one used in the balanced mode.)

SOLUTION: Identical heat exchangers have the same factor Ah. Since $(c_p \dot{m})_{min}$ is supposed to be the same for both, they also have the same NTU (see Equation 221 b).

Consider the limits of strongly unbalanced and nearly balanced modes:

a) Strongly unbalanced: $C^* \ll 1$

$$\left.\begin{aligned} &\text{Eq. 220:} \quad \varepsilon_u \approx 1 - \exp(-NTU) \\ &\text{Eq. 222:} \quad \varepsilon_b = \frac{1}{1 + 1/NTU} \end{aligned}\right\} \quad \varepsilon_u > \varepsilon_b$$

b) Nearly balanced: $C^* \approx 1$. Naturally, in this case, the unbalanced mode should approach the performance of the balanced one; because of $C^* \approx 1$ we have $NTU(1-C^*) \ll 1$:

$$\text{Eq. 220:} \quad \varepsilon_u \approx \frac{1-(1-NTU(1-C^*))}{1-C^*(1-NTU(1-C^*))} = \frac{NTU(1-C^*)}{1-C^* + C^* NTU(1-C^*)}$$

$$= \frac{NTU}{1 + C^* NTU} \approx \varepsilon_b$$

■

As we have found in examples in Chapter 3, the entropy generated in the Earth's interior cannot be transported by conduction because this mode of heat transfer is not effective enough under the given circumstances. It is therefore assumed that heat is transferred convectively, and the material of the mantle is modeled as an ideal fluid as discussed in the Interlude. Make a model of convective motion of the Earth's mantle in which a blob of matter rises from the interior to the surface adiabatically (as in the model for the Earth's atmosphere in Section 3.3.6, or as discussed in Section 4.6.3). a) Assume that the heating at a given radius is due to entropy production as the result of radioactive decay (which is taken to be distributed evenly in the entire mantle) at smaller radii, and friction which assumes all the energy released by the heat engine represented by the convective motion to be dissipated whereby more entropy is returned to the rising matter. Show that in this case the entropy current entering a thin shell at radius r and driving the heat engine is given by

$$I_s(r) = \frac{1}{T(r)} \frac{m(r) - m(r_c)}{m(R) - m(r_c)} I_E(R)$$

where r_c and R are the radius of the core (bordering on the mantle) and the radius of the Earth, respectively. $I_E(R)$ is the energy flux penetrating the Earth's surface from the interior of the planet. b) It is found that the gravitational field g is roughly constant in the entire mantle. Show that in this case

$$I_s(r) = \frac{1}{T(r)} \frac{r^2 - r_c^2}{R^2 - r_c^2} I_E(R)$$

c) Show that if you assume the temperature gradient through the mantle to be adiabatic, it can be expressed by

$$\mathcal{P} \approx g \cdot I_E(R) \int_{r_c}^{R} \frac{r^2 - r_c^2}{R^2 - r_c^2} \gamma^* \rho \cdot \kappa_s dr$$

where γ^* is the Grüneisen ratio defined in Problem 24 of the Interlude, and κ_s is the adiabatic compressibility of the material. (*Hint:* See Problem 24 of Chapter 1 and Problem 27 in the Interlude.) d) The values of the different physical parameter of the Earth's interior have to be derived from seismic and other measurements.[5] Take an average density of 4500 kg/m³, and average values of 10 N/kg, 0.8, and $3 \cdot 10^{-12}$ /Pa for the gravitational field, the Grüneisen ratio, and the adiabatic compressibility, respectively. The lower boundary of the mantle is at 3500 km from the center of the Earth, and the energy flux through the surface is $31 \cdot 10^{12}$ W. Estimate the efficiency of the convective motion interpreted as a heat engine.

See Stacey (1992) for numerical values for the Earth's interior.

SOLUTION:

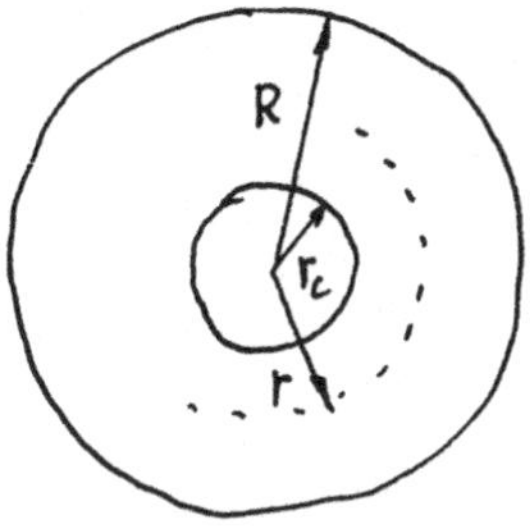

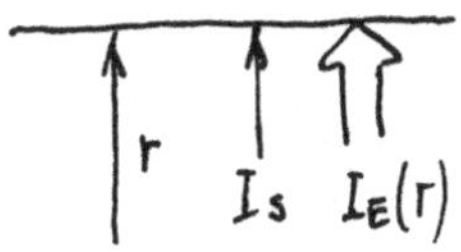

a) $$I_s(r) = \frac{1}{T(r)} I_E(r)$$

Steady-state balance of energy for mantle from r_c to r:

$$I_E(r) = \Sigma_E$$

where $\Sigma_E = \sigma_E \, m_{r_c \to r} = \sigma_E \left(m(r) - m(r_c) \right)$

Since $I_E(R) = \sigma_E \left(m(R) - m(r_c) \right)$

$$\to \quad I_E(r) = \frac{m(r) - m(r_c)}{m(R) - m(r_c)} I_E(R)$$

$$\to \quad I_s(r) = \frac{1}{T(r)} \frac{m(r) - m(r_c)}{m(R) - m(r_c)} I_E(R)$$

b) $g(r) = G\, m(r)/r^2 = \text{const} \quad \to \quad m(r) = \text{const} \cdot r^2$

$$\to \quad I_s(r) = \frac{1}{T(r)} \frac{r^2 - r_c^2}{R^2 - r_c^2} I_E(R)$$

c) Chapter 1, Problem 24:

$$\mathcal{P} = \int_{T(r_c)}^{T(R)} I_s(r)\, dT = \int_{T(r_c)}^{T(R)} I_E(R) \frac{r^2 - r_c^2}{R^2 - r_c^2} \frac{dT}{T}$$

adiabatic temperature gradient: $$\mathcal{P} = I_E(R) \int_{r_c}^{R} \frac{r^2 - r_c^2}{R^2 - r_c^2} \left.\frac{dT}{dr}\right|_{ad} dr$$

Interlude, Problem 27: $\left.\frac{dT}{dr}\right|_{ad} = -\gamma^* T \rho g \kappa_s$

$$\rightarrow |\mathcal{P}| \approx g\, I_E(R) \int_{r_c}^{R} \frac{r^2 - r_c^2}{R^2 - r_c^2}\, \gamma^* \rho\, \kappa_s\, dr$$

d) Numerical values:

$$|\mathcal{P}| \approx 10 \cdot 31 \cdot 10^{12}\, \frac{0.8 \cdot 4500 \cdot 3 \cdot 10^{-12}}{(6.4 \cdot 10^6)^2 - (3.5 \cdot 10^6)^2} \int_{r_c}^{R} (r^2 - r_c^2)\, dr \;\; W$$

$$= 1.2 \cdot 10^{-7} \left(\frac{1}{3} r^3 - r_c^2 r\right)_{r_c}^{R} \;\; W$$

$$= 1.2 \cdot 10^{-7} \left(\frac{1}{3} R^3 - r_c^2 R - \frac{1}{3} r_c^3 + r_c^3\right) \;\; W$$

$$= 1.2 \cdot 10^{-7} \left(\frac{1}{3} (6.4 \cdot 10^6)^3 - (3.5 \cdot 10^6)^2 6.4 \cdot 10^6 + \frac{2}{3} (3.5 \cdot 10^6)^3\right) \;\; W$$

$$|\mathcal{P}| \approx 4.4 \cdot 10^{12}\; W$$

If we take $0.5 \cdot I_E(R)$ for the energy flux which is heating the convective heat engine, we get an estimate of 30% for its efficiency.

■

EPILOGUE

Solutions of Selected Problems

Consider a uniform viscous fluid such as the one introduced in Section E.1.1. Let the fluid have a pressure, and entropy capacity, and a latent entropy like the ideal gas, but let friction be present. a) Show that the energy added in heating of the fluid at constant volume is

$$Q = E(V, T_f) - E(V, T_i)$$

where the indices i and f refer to the initial and the final states, respectively. b) Consider heating at constant pressure. Show that in this case the energy added in heating can be expressed as

$$Q = H(P, T_f) - H(P, T_i) + \int_{t_i}^{t_f} a\dot{V}^2 dt$$

Show that this quantity is always less than the difference of the enthalpies at the end and the beginning. c) Assume the friction factor a, the rate of change of the volume, and the energy current in heating to be given. Take the rate of change of the volume to be constant. Show that the initial value problem of the fluid takes the form

$$\frac{C_V}{T}\dot{T} - \frac{1}{T}I_E(t) + \frac{a}{T}\dot{V}^2 = -\frac{nR}{V_o + \dot{V}t}\dot{V}$$

$$T(t = 0) = T_o$$

where V_o is the volume at $t = 0$.

SOLUTION:

a) According to definition, $Q = -\int I_E \, dt$. With Equation (21) we have $Q = -\int T I_s \, dt$.

$V = \text{const}$: $\quad Q = -\int T I_s \, dt = -\int T\left[-\Lambda_V \dot{V} - K_V \dot{T} - \frac{1}{T} a \dot{V}^2\right] dt$

↑ Equation (26)

$$= \int_{T_i}^{T_f} T K_V \, dT = \int_{T_i}^{T_f} \frac{\partial E(V,T)}{\partial T} dT$$

$$= E(V, T_f) - E(V, T_i)$$

b) $P = \text{const}$:
$$Q = -\int T I_s \, dt = -\int T(-\dot{S} + \Pi_s)\, dt$$
$$= \int (T\dot{S} + a\dot{V}^2)\, dt$$
↑ Equation (22)
$$= \int (T\Lambda_p \dot{P} + T K_p \dot{T})\, dt + \int a\dot{V}^2 dt$$
$$= \int_{T_i}^{T_f} C_p \, dT + \int a\dot{V}^2 dt$$
$$= H(P, T_f) - H(P, T_i) + \int_{t_i}^{t_f} a\dot{V}^2 dt$$

Since $a \le 0$: $Q \le H_f - H_i$

c) Combining Equations (26) and (21) we obtain:

$$\Lambda_V \dot{V} + K_V \dot{T} = \frac{1}{T} I_E - \frac{a}{T}\dot{V}^2$$

$\dot{V} = \text{const} \implies V(t) = V_0 + \dot{V}t$

Since $\Lambda_V = nR/V$, and $K_V = C_V/T$ (as in Chapter 2):

$$\frac{nR}{V_0 + \dot{V}t}\dot{V} + \frac{C_V}{T}\dot{T} = \frac{1}{T} I_E - \frac{a}{T}\dot{V}^2 .$$

■

Consider the uniform viscous fluid described in Problem 1. Show that for isothermal heating at a temperature T, with prescribed entropy current $I_s(t)$, the differential equation for the volume becomes

$$\frac{a}{T}\dot{V}^2 + \frac{nR}{V}\dot{V} = -I_s(t)$$

SOLUTION:

Balance of entropy: $\dot{S} = -I_s + \Pi_s$

where $\dot{S} = \Lambda_V \dot{V} + K_V \dot{T}$

$\Pi_s = -a/T\,\dot{V}^2$ (Equation 22)

$T =$ const: $\Lambda_V \dot{V} = -I_s + \left(-\frac{a}{T}\dot{V}^2\right)$

$$\Lambda_V \dot{V} + \frac{a}{T}\dot{V}^2 = -I_s$$

With $\Lambda_V = nR/V$: $\frac{a}{T}\dot{V}^2 + \frac{nR}{V}\dot{V} = -I_s(t)$

■

Consider a pebble bed through which air is pumped. (Pebble bed heat storage is one of the means of storing energy from the sun for heating purposes.) Show that in the purely one-dimensional case the differential equations for the temperature of the air and of the pebbles as a function of time and of axial position are given by

$$(\rho \cdot c_p)_a e \frac{\partial T_a}{\partial t} = -\frac{1}{A}(c_p I_m)_a \frac{\partial T_a}{\partial x} - h^*(T_a - T_p)$$

$$(\rho \cdot c_p)_a (1-e) \frac{\partial T_p}{\partial t} = h^*(T_a - T_p)$$

Here, A is the cross section of the pebble bed, while e denotes the bed void fraction. h^* is the heat transfer coefficient between air and pebbles multiplied by the pebble surface per unit bed volume. The following additional assumptions have been made: no heat loss to the environment and no temperature gradient within the pebbles.

SOLUTION:

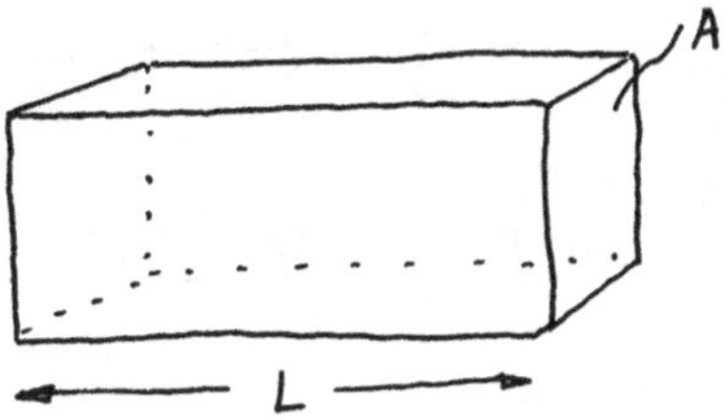

$V = A \cdot L$

$V_a = e A L$ air

$V_p = (1-e) A L$ pebbles

For the air, we start with the equations of balance given in (130) and (131). Because of (130a) we have

Mass: $$\frac{\partial \rho}{\partial t} + \frac{\partial \rho v}{\partial x} = 0$$

Momentum: $$\rho \frac{\partial v}{\partial t} + \rho v \frac{\partial v}{\partial x} + \frac{\partial P}{\partial x} = 0$$

Energy: $$\rho \frac{\partial u}{\partial t} + \rho v \frac{\partial u}{\partial x} + v \frac{\partial P}{\partial x} + P \frac{\partial v}{\partial x} = \sigma_E$$

There are three changes compared to the general equations in the text. First of all, they have been applied to the one-dimensional case. Second, the momentum current density (stress) of non-viscous air is the pressure P. Third, a source rate density of energy has been added to the balance of energy, since we model the effects of transfer of entropy (and energy) from air to pepples as a source (or sink) inside the volume.

Now we introduce the specific enthalpy $h = u + \frac{1}{\rho}P$ in the energy equation:

$$\rho\frac{\partial h}{\partial t} + \rho v\frac{\partial h}{\partial x} - \frac{\partial P}{\partial t} + \frac{1}{\rho}P\frac{\partial \rho}{\partial t} - v\frac{\partial P}{\partial x} + \frac{1}{\rho}Pv\frac{\partial \rho}{\partial x} + v\frac{\partial P}{\partial x} + P\frac{\partial v}{\partial x} = \sigma_E$$

$$\rho\frac{\partial h}{\partial t} + \rho v\frac{\partial h}{\partial x} \underbrace{- \frac{\partial P}{\partial t} + \frac{1}{\rho}P\frac{\partial \rho}{\partial t} + \frac{1}{\rho}Pv\frac{\partial \rho}{\partial x} + P\frac{\partial v}{\partial x}} = \sigma_E$$

$$\text{balance of mass:} \quad -\frac{\partial P}{\partial t} + \frac{P}{\rho}\left(-\rho\frac{\partial v}{\partial x}\right) + P\frac{\partial v}{\partial x}$$

Since $P = \text{const}$, we also have $\partial P/\partial t = 0$. Therefore:

$$\rho\frac{\partial h}{\partial t} + \rho v\frac{\partial h}{\partial x} = \sigma_E$$

σ_E is the rate of transfer of energy from air to pebbles per volume of air. If h is the heat transfer coefficient, and A^* is the surface area of the of the pebbles surrounded by air, we have

$$\sigma_E = -\frac{A^* h}{e A L}(T_a - T_p)$$

With $= c_p(T_a - T_{ref})$, we obtain

$$e(\rho c_p)_a \frac{\partial T_a}{\partial t} + e(\rho v c_p)_a \frac{\partial T_a}{\partial x} = -\frac{A^* h}{A L}(T_a - T_p)$$

The flux of mass of air through the bed is $I_m = eAv\rho$ which yields the final result.

For the pebbles, we use an extended version of Equation (130) in Chapter (3):

$$(\rho c)_p \frac{\partial T_p}{\partial t} = \frac{\partial}{\partial x}\left(k_E \frac{\partial T}{\partial x}\right) + \sigma_E \qquad \text{with } k_E = 0$$

Here, σ_E is calculated with respect to the volume of the pebbles:

$$\sigma_E = \frac{A^* h}{(1-e) A L}(T_a - T_p)$$

Therefore we have:

$$(1-e)(\rho c_p)_p \frac{\partial T_p}{\partial t} = \frac{A^* h}{A L}(T_a - T_p)$$

■